United States Office of Water (4303T)
EPA-821-R-04-013
Environmental Protection Agency
Washington, DC 20460

June 2004

EPA-821-R-04-013

Economic and Environmental Benefits Analysis of the Final Effluent Limitations Guidelines and New Source Performance Standards for the Concentrated Aquatic Animal Production Industry Point Source Category

Stephen L. Johnson
Acting Deputy Administrator

Benjamin H. Grumbles
Acting Assistant Administrator, Office of Water

Mary T. Smith
Director, Engineering and Analysis Division

Christopher J. Miller
Lisa McGuire
Renee Selinsky Johnson
Project Analysts

Engineering and Analysis Division
Office of Science and Technology
U.S. Environmental Protection Agency
Washington, D.C. 20460

June 2004

ACKNOWLEDGMENTS AND DISCLAIMER

The Office of Science and Technology prepared this document with the support of Eastern Research Group, Incorporated and Tetra Tech, Incorporated.

CONTENTS

TABLES

FIGURES

EXECUTIVE SUMMARY

ES.1   INTRODUCTION

The U.S. Environmental Protection Agency (EPA) has finalized Clean Water Act effluent limitations guidelines and new source performance standards for aquatic animal production facilities. The regulation establishes for the first time technology-based effluent guidelines for wastewater discharges from new and existing aquatic animal production facilities that discharge directly to U.S. waters. This document summarizes the costs, economic impacts, and the environmental benefits associated with this regulation.

EPA's National Pollutant Discharge Elimination System (NPDES) regulations define when a hatchery, fish farm, or other facility is a concentrated aquatic animal production facility that is a point source subject to the NPDES permit program. See 40 CFR 122.4. In defining "concentrated aquatic animal production (CAAP) facility," the NPDES regulations distinguish between warmwater and coldwater species of fish and defines a CAAP by, among other things, the size of the operation and frequency of discharge. A facility is a CAAP if it meets the criteria in 40 CFR 122 Appendix C or if it is designated as a CAAP by the NPDES program director on a case-by-case basis. For more information, see the preamble of the final regulation.

Aquatic animals raised for commercial and noncommercial purposes are diverse, ranging from species produced for human consumption as food to species raised for recreational purposes. The animals may be raised in a variety of production systems. The choice of a production system is influenced by a variety of factors including species, economics of production, markets, local water resources, land availability, and operator preference. Some production systems, especially those needed to produce species intended for release into the wild or other natural environments, are designed to provide a suitable environment that imitates the natural environment of the species.

Entities potentially regulated by this action include facilities engaged in concentrated aquatic animal production, which may include both commercial (for profit) and noncommercial (public) facilities. By North American Industry Classification System (NAICS), regulated entities include "Finfish Farming and Fish Hatcheries" (NAICS 112511) and "Other Animal Aquaculture" (NAICS 112519).

On December 29, 2003, the Office of the Federal Register published a Notice of Data Availability (USEPA, 2003; 68 FR 75068). In the Notice, EPA summarized the data received since the proposed rule and described how the Agency might use the data for the final rule. The Notice discussed EPA's detailed survey effort. This second phase of data collection involved mailing a survey that asked for more detailed and specific information than the initial screener survey. The detailed survey was a stratified sample population of facilities identified from the screener survey. EPA received responses from 205 facilities. The surveyed population included a statistically representative sample of facilities that reported producing aquatic animals with flow through, recirculating and net pen systems. EPA also surveyed a small number of facilities that would not have been subject to the proposed requirements.

This Economic and Environmental Benefit Analysis (EEBA) summarizes EPA's analysis of the estimated annual compliance costs and the economic impacts that may be incurred by affected operations that are subject to the final rule. The report covers financial impacts to regulated aquaculture facilities, along with market and other secondary impacts such as impacts on prices, quantities, trade, employment, and output. This report also present EPA's estimates of the environmental benefits associated with the final regulation. It also responds to requirements for small business analyses under the Regulatory Flexibility Act (RFA) as amended by the Small Business Regulatory Enforcement Fairness Act (SBREFA) and for cost-benefit analyses under Executive Order 12866 and the Unfunded Mandates Reform Act (UMRA).

Additional information on EPA's costing methodology and estimated costs are provided in the *"Technical Development Document for the Final Effluent Limitations Guidelines and New Source Performance Standards for the Concentrated Aquatic Animal Production Point Source Category"* [EPA-821-R-04-012] referred to in this report as the "Development Document." That document presents the technical information that formed the basis for EPA's decisions in the final rule. It also describes, among other things, the data collection activities, the wastewater treatment technology options considered by the Agency as the basis for effluent limitations guidelines and standards, the pollutants found in wastewaters, the estimates of pollutant removals associated with certain pollutant control options, and the cost estimates related to reducing the pollutants with those technology options. The Proposal EIA provides a detailed industry profile of the U.S. aquaculture industry (USEPA, 2002b, Section 2).

ES.2   DATA AND METHODOLOGY

ES.2.1   Data Sources

EPA's economic analysis relied on a wide variety of data and information sources. Data sources used in the economic analysis include EPA's Screener Questionnaire and EPA's Detailed Questionnaire for the Aquatic Animal Production Industry, data from the U.S. Department of Agriculture (USDA) and the Joint Subcommittee on Aquaculture (JSA),[1] as well as the academic literature and industry journals.

EPA collected facility-level production data from individual aquatic animal producers through a screener survey administered under the authority of the CWA Section 308 (USEPA, 2001). EPA used response data from the screener survey to classify and subcategorize facilities by production method, species produced and production level, and water treatment practices in place prior to the proposed regulation. EPA used the information from the screener survey to identify a subset of facilities to receive the detailed questionnaire. Details about the use of screener survey data are in the Technical Development document for the proposed rule (USEPA, 2002c).

Like the screener survey, EPA administered the detailed survey under the authority of the CWA Section 308 (USEPA, 2002e). EPA used response data from the survey to classify and subcategorize facilities by production method, species produced and production level, and water treatment practices in place prior to the proposed regulation. For commercial operations, the survey instrument collected financial and economic information at the aquaculture enterprise, the facility, and the company that owned the facility. For public or noncommercial operations, EPA collected financial and economic information on operating costs and funding sources.

---

[1] JSA is a Federal interagency coordinating group, authorized under the National Aquaculture Act of 1980 and the National Aquaculture Improvement Act of 1985.

More information on the sources of data used for this analysis is provided in Section 1.2 of this report, the rulemaking preamble, and in EPA's *Development Document* (USEPA, 2004).

ES.2.2 Regulated Community

EPA estimates that a total of 242 facilities will be in scope of this final regulation. Table ES-1 summarizes the estimated number and type of facilities affected by the rule, based on the production threshold of 100,000 lb/year. These 242 facilities consist of 101 commercial facilities and 141 noncommercial facilities (Federal, State, Tribal, and Alaska nonprofit organizations). Of the 101 commercial facilities that potentially incur costs under the final rule, 82 are in the Flow Through and Recirculating Subcategory. Of the commercial facilities represented in the detailed summary, 69 are projected to show long-term profitability (estimated by cash flow) prior to the final rule. The remaining 32 facilities are termed "baseline closures" because they are projected to show long-term unprofitability in absence of the final rule. EPA has identified no academic/research facility in the detailed questionnaire that produced more than 100,000 lbs/yr. See Section 4.3 of this report for more information on EPA's cost estimates of this final regulation.

ES.2.3 Cost Methodology

Costs associated with regulatory compliance are used to estimate the economic impact of the effluent limitations guidelines and standards on the aquaculture industry. Economic impacts are a function of the estimated costs of compliance to achieve the requirements, which may include initial fixed and capital costs, and annual operating and maintenance (O&M) costs. EPA's estimation of these costs began by identifying the practices and technologies that could be used as a basis to meet particular requirements. EPA estimated compliance costs for each facility, based on the implementation of the practices or technologies that minimizes the cost to meet particular requirements.

Table ES-1
Estimated Number of Affected Facilities with Production > 100,000 lbs/yr

Organization	Estimated Number of Facilities[1]		
	Baseline Closures[2]	Not Baseline Closures[3]	Total
Commercial	32 (28)	69 (69)	101 (97)
Noncommercial	NA (NA)	141 (141)	141 (141)
TOTAL	32 (28)	210 (210)	242 (238)

Source: Estimated by USEPA.
Source: EPA estimates from detailed survey (USEPA, 2002e).
NA: EPA does not analyze closures for government facilities.
[1] Numbers in (parentheses) are facilities EPA projects are *not* currently achieving the requirements of the final rule.
[2] Projected baseline closures are estimated using cash flow analysis. When net income analysis is assumed for earnings, the number of commercial baseline closures increases to 43. Baseline closures would not be projected to incur impacts from a new rule in accordance with EPA's Guidelines for Preparing Economic Analyses (USEPA, 2000). Baseline closures (based on cash flow) are therefore not included in estimates of costs for this rule.
[3] Total costs and economic impacts for this rule are estimated using incremental compliance costs incurred by the facilities that are not baseline closures and not in compliance with the rule at time of final signature (i.e., 238 facilities are expected to incur costs under this rule: 97 commercial and 141 noncommercial facilities).

EPA developed cost estimates for capital, one-time fixed, land, and annual O&M costs for the implementation and use of the different best management practices and treatment technologies targeted under the regulatory options considered in the Final Rule. EPA developed the cost estimates from information collected during the detailed survey, site visits, sampling events, published information, vendor contacts, industry comments, and engineering judgment. Additional information on how EPA developed the cost models is provided in the *Development Document* (USEPA, 2004). See also EPA's detailed responses to public comments received on proposal and EPA's Notice on the proposed rule. These comments and the Agency's response are in the Comment Response Document that is available in the rulemaking record.

EPA initially estimates compliance costs in 2001 dollars. These costs are translated to 2003 dollars using a Construction Cost Index (ENR, 2004). Total costs for this rule are therefore given in 2003 dollars, but costs used for impact analysis are maintained in 2001 dollars to remain consistent with 2001 revenues in detailed survey responses.

ES.2.4   Economic Impact Methodology

For this final regulation, EPA evaluates the economic impacts on both new and existing commercial and also noncommercial operations. The following is a description of the approach EPA uses to prepare these analyses.

Existing Commercial Facilities

EPA uses several measures to evaluate possible impacts on existing commercial facilities. These measures examine the possibility of business closure and corresponding direct impacts on employment and communities and indirect and national impacts associated with closures.

To evaluate impacts to commercial facilities, EPA conducts a closure analysis that compares projected earnings, with and without cost of compliance with the final regulation, for the period from 2005 to 2015. EPA uses two measures to estimate earnings for the purposes of its closure analysis: cash flow and net income. The difference between the cash flow and net income calculations is depreciation (a non-cash cost). Depreciation is included as a cost in the net income basis but not in the cash flow basis.

Analysis using net income is more likely to identify baseline closures and could demonstrate additional regulatory closures associated with the rule. All other analytical results (for example other measures of economic impacts, costs) presented in this final action reflect discounted cash flow as the basis for earnings; EPA's analyses indicate that use of net income will not materially change results.

For this analysis, EPA calculates the difference between gross revenues and total expenses reported in the detailed questionnaire and reduced the value by the estimated Federal and State taxes to calculate net income. EPA then adds the non-cash expense of depreciation (when it was reported in the questionnaire) to net income to calculate cash flow. This approach is consistent with the guidance from the Farm Financial Standards Council (FFSC, 1997) and several business financial references (Brigham and Gapenski, 1997; Jarnagin, 1996; and Brealy and Myers, 1996). As part of this analysis, EPA examines the possibility of closure under three forecasting methods to project future earnings.

Baseline closures should not be attributed to the rule, but rather should be classified as baseline closures (USEPA, 2000). EPA did not analyze facilities with negative net earnings, under two or three of the forecasting methods before they incur pollution control costs (i.e., baseline closures). EPA determined that 32 out of 101 commercial facilities are baseline closures. Given that no closures are projected to occur under the final rule, there are no employment and other direct and indirect impacts estimated for this rule. EPA also performs additional sensitivity analysis on these results; a total of 43 baseline closures are projected under net income analysis.

In addition to its closure analysis, EPA also prepared additional analyses to assess other potential effects, including an analysis of additional moderate impacts using a sales test, an evaluation of financial health using an approach similar to that used by USDA, and an assessment of possible impacts on borrowing capacity.

For the purposes of assessing economic achievability, EPA assumes that facilities are unable to expand production to cover the cost of the rule and also cannot pass costs on to consumers. The facility, therefore, must absorb all increased costs. If it cannot do so and must remain in operation, all production is assumed lost. More information on EPA's rationale for this approach is provided in Section 3.6. EPA's assumption of no cost pass through is a more conservative approach to evaluating economic achievability among regulated entities. (To evaluate market and trade level impacts, however, EPA assumes all costs are shifted onto the broader market level as a way of assessing the upper bound of potential effect; see Section 5.4.)

See Section 3.2 for more information on EPA's approach for addressing economic impacts at regulated commercial facilities.

Noncommercial Facilities

For the final rule, EPA collected information on how U.S. Fish and Wildlife Service (FWS) and State agencies make decisions about operating or closing public hatcheries. EPA confirmed that public hatcheries close; the FWS hatchery system once had as many as 250 hatcheries and it now operates fewer than 90 facilities. Closures may result from funding cuts (e.g., Mitchell Act funds and the Willard National Fish Hatchery or General Funds for State hatcheries) or revision of a program's mission and goals (e.g., increase focus on endangered species versus provision of recreational services). Closures may also result from water quality impacts associated with aquaculture activities. The costs of upgrading pollution control at public hatcheries are not the primary reason for closure, but costs may tip the balance of a particular hatchery toward a closure decision.

In the absence of well defined tests for projecting public facility closures, EPA compares pre-tax annualized compliance costs (in 2001 dollars) to 2001 operating budgets for noncommercial facilities including State, Federal, and Tribal facilities ("Budget Test"). For the purposes of this analysis, EPA assumes a 5 percent and 10 percent threshold value as an indicator of potential financial impacts at noncommercial facilities. Accordingly, costs exceeding 5 percent and 10 percent signal potential "moderate" and "adverse" financial impacts, respectively. For Alaska nonprofit facilities, impacts are estimated by comparing pre tax annualized costs to harvest revenues.

Impacts to noncommercial facilities are expected to be a function of a facility's ability to access additional funds from user fees. As part of analyses, EPA examines the ability of State-owned hatcheries to recoup compliance costs through increases in funding derived solely from user fees. All States and the District of Columbia have fishing license fees for residents. The license fees are not raised every year even though costs increase through inflation. Instead, when fees are raised or a fish stamp instituted, the raise or new fee is usually a round number such as $3, $5, or $10. A $3 to $5 hike in State fishing license fees translates into an increase in fees of about 20 percent to 35 percent. Although all States report having fishing license fees, if a state hatchery reports no funding from user fee sources, EPA considers that facility to be unable to recoup increased costs through increased funding from user fees.

See Section 3.3 for more information on EPA's approach for addressing economic impacts at regulated noncommercial facilities.

New Commercial Facilities

To assess effects on new businesses, EPA's analysis considers the barrier that new compliance costs may pose to entry into the industry for a new facility. In general, it is less costly to incorporate waste water treatment technologies as a facility is built than it is to retrofit existing facilities. Therefore, where a rule is economically achievable for existing facilities, it will also be economically achievable for new facilities that can meet the same guidelines at lower cost. Similarly, even where the cost of compliance with a given technology is not economically achievable for an existing source, such technology may be less costly for new sources and thus have economically sustainable costs. It is possible, on the other hand, that to the extent the up-front costs of building a new facility are significantly increased as a result of the rule, prospective builders may face difficulties in raising additional capital. This could present a barrier to entry. Therefore, as part of its analysis of new source standards, EPA evaluates barriers to entry. If the requirements promulgated in the final regulation do not give existing operators a cost advantage over new source operators, then EPA assumes new source performance standards do not present a barrier to entry for new facilities. See Section 3.5 for more information.

EPA's analysis includes all commercial facilities within scope of the rule, including those that are baseline closures. EPA examines the (1) proportion of commercial facilities that incur no costs, (2) proportion of commercial facilities that incur no land or capital costs, and (3) ratio of incremental land and capital costs to total company assets. The cost to asset ratio is calculated using company data because asset data were collected only at the company level. EPA calculates the ratio for each company and use the average of the ratios, rather than taking the ratio of average debt to average assets.

ES.3   EPA'S ESTIMATE OF REGULATORY COSTS

ES.3.1   Costs to Regulated Facilities

EPA estimates the annual incremental costs of compliance using the capital and recurring costs derived in the *Development Document* (USEPA, 2004). EPA converts these costs to incremental annualized costs. Annualized costs better describe the actual compliance costs that a regulated aquaculture facility would incur, allowing for the effects of interest, depreciation, and taxes. EPA uses these annualized costs to estimate the total annual compliance costs and to assess the economic impacts of the final requirements to each regulated operation. Section 3.1 provides more details on EPA's cost annualization model and methodology.

The final option sets narrative standards for the control of solids based on implementation through operational measures addressing (1) feed management, (2) cleaning and maintenance, (3) storage of feed, drugs and pesticides to prevent spills, (4) record keeping on feed, cleaning inspections, maintenance, and repairs.

Table ES-2 summarizes the total national costs for the final regulation. Estimated annualized cost for the final regulation is $1.4 million (2003 dollars). Noncommercial facilities account for about 80 percent of the total cost of the rule. This estimated total cost reflects aggregate compliance costs incurred by facilities that produce more than of 100,000 lb/year and will be affected by this final regulation.

These aggregated cost estimates reflect pre-tax costs. However, EPA's model calculates both pre-tax and post-tax costs. The post-tax costs reflect the fact that a commercial regulated operation would be able to depreciate or expense these costs, thereby generating a tax savings. Post-tax costs thus are the actual costs the regulated facility would face. Post-tax costs are also used to evaluate impacts on regulated facilities using a discounted cash flow and net income analysis. Pre-tax costs reflect the estimated total social cost of the regulations, including lost tax revenue to governments. Pre-tax dollars are used when comparing estimated costs to monetized benefits that are estimated to accrue under the final regulations (see Sections 7 and 8 of this report). Estimated costs have been converted from 2001 dollars to 2003 dollars using the Construction Cost Index (ENR, 2004).

Table ES-2
National Costs: Total by Subcategory and Option

Production System[1]	Owner	Pre-tax Annualized costs ($000, 2003 Dollars)
		Final Option
Flow Through and Recirculating	Commercial	$256
	Noncommercial	$1,149
Netpen	Commercial	$36
	Noncommercial	$0
Total Pre-tax[2]		$1,442

Note: May not sum due to rounding
1. Costs exclude baseline closure facilities; see Table ES-1.
2. Total annual post-tax cost is $1,362 for the final option. Costs are calculated over the 2005 to 2015 period with a 7percent real discount rate.

ES.3.2 Costs to the Permitting Authority (States and Federal Governments)

All of the aquaculture facilities in the scope of the final rule are currently permitted, so incremental administrative costs of the regulation to the permitting authority are expected to be negligible. However, Federal and State permitting authorities will incur a burden for tasks such as reviewing and certifying the BMP plan and reports on the use of drugs and chemicals. EPA estimates these costs to be $13,176 for the three-year period covered by EPA's information collection request, or roughly $4,392 per year. These results show that the recordkeeping and reporting burden to the permitting authorities is less than two-tenths of one percent of the pre-tax compliance cost for the final rule.

ES.4 EPA'S ESTIMATE OF REGULATORY IMPACTS

ES.4.1 Financial Effects to Regulated Operations

This section describes the results of EPA's economic analysis of the effects of this final regulation on both new and existing commercial and also noncommercial operations. Based on the results of this and other analyses, EPA concludes that effluent requirements under the final option under this rule is economically achievable. See Chapter 5 of this report for more information.

For the purposes of this analysis, EPA assumes these operations are not able to pass on the compliance costs due to the regulation. EPA's assumption of "no cost pass through" is a more conservative approach to evaluating economic achievability among regulated entities. See Section 3.6 of this report. (To evaluate market and trade level impacts, however, EPA assumes all costs are shifted onto the broader market level as a way of assessing the upper bound of potential effect; see Section 5.3)

Existing Commercial Facilities

Table ES-3 shows the effects on commercial operations from the final regulation based on EPA's economic analysis. As shown, EPA projects no enterprise or facility closures as a result of the final rule under the cash flow assumptions for earnings. The Agency therefore considers the final rule to be economically achievable for commercial facilities (and companies). For more information see Section 5.1 of this report.

EPA expects some operations will incur additional moderate impacts, based on an analysis that shows that some operations will incur compliance costs in excess of 5 percent of annual revenue. For the final regulation, 4 commercial facilities incur costs greater than 5 percent of sales, affecting about 4 percent of all existing regulated facilities in the continuous discharge subcategory and approximately 6 percent of all existing regulated facilities that are not projected to be baseline closures. No commercial facilities have costs that exceed 10 percent of annual revenue. EPA's analysis also shows one company potentially experience an impact on borrowing capacity. EPA considers these as "moderate" impacts (Section 3.2). EPA's analysis also shows no expected change in financial health for any of the commercial facilities as a result of the final regulation. This is based on EPA evaluation of the companies represented in the Agency's detailed questionnaire.

Noncommercial Facilities

Table ES-3 shows the impacts on noncommercial operations from the final regulation based on EPA's economic analysis. For the final option, 4 facilities incur costs exceeding 10 percent of budget. EPA assumes that those facilities that face costs exceeding 10 percent of their budget would be adversely affected by the final regulation. None of these facilities report user fee funds; EPA could not conduct additional supplemental analyses to determine whether an increase in fees could offset these results. EPA's results, therefore, indicate that 3 percent of all noncommercial operations may be adversely affected by this final regulation. These operations may be vulnerable to closure based on the results of the Agency's budget test but constitute a relatively small percent of the population.

Under a 5 percent budget test, 8 facilities exceed the threshold under the final regulation. Among facilities that experience an increase in costs exceeding 5 percent, EPA assumes these facilities would face moderate financial impacts but would not be adversely affected. These results show that an additional 6 percent of all noncommercial operations (not counting those adversely affected) would experience some moderate impact associated with the costs of the rule. Some of these facilities report user fees revenues. Therefore, EPA conducts additional supplemental analyses to determine whether an increase in user fees could offset these results (see Section 5.1.2.2).

Given that the results of EPA's analysis projects that a small share of regulated noncommercial facilities may incur costs exceeding 10 percent of budget, estimated at 3 percent of facilities, the Agency considers these final technology options to be economically achievable for noncommercial facilities. For more information see Section 5.1 of this report.

Table ES-3
Economic Effects: Existing Commercial & Noncommercial Operations

Threshold Test	Estimated Number of In-Scope Facilities	Final Option
Commercial Operations		
Closure Analysis[1]	**101**	**0**
Sales test >3%	101	4
Sales test >5%	101	4
Sales test >10%	101	0
Change in Financial Health	NA[2]	0
Credit test >80%	NA[2]	1
Noncommercial Facilities[5]		
Budget test >3% (all facilities)	**141**	**19**
State owned only (# with user fees)[4]	106	12 (8)
Federal owned only	33	7
Alaskan Non-Profit[3]	2	0
Budget test >5% (all facilities)	**141**	**12**
State owned only (# with user fees)[4]	106	8 (8)
Federal owned only	33	4
Alaskan Non-Profit[3]	2	0
Budget test >10% (all facilities)	**141**	**4**
State owned only (# with user fees)[4]	106	0 (0)
Federal owned only	33	4
Alaskan Non-Profit[3]	2	0

Source: Estimated by USEPA using results from facility-specific detailed questionnaire responses, see Chapter 3.
1) Closure analysis assumes discounted cash flow for earnings. A total of 32 facilities are projected to be baseline closures; these facilities cannot be attributed to this rule.
2) Analysis performed at the company level. EPA evaluated 34 unweighted companies representing the 101 weighted facilities from the detailed questionnaire. The statistical weights, however, are developed on the basis of facility characteristics and therefore cannot be used for estimating the number of companies.
3) Two Alaska non-profit organizations are within the scope of this rule, but did not receive a detailed survey. They were costed using screener survey data. Economic impacts were calculated using publically available information.
4) Some State-owned facilities reported that they relied, in part, on funds from State user fee operations. These numbers are reported in parenthesis and are included in the overall numbers as well.

5) EPA maintains that there is potential for Tribal facilities to be present within the population of noncommercial facilities affected by this rule, despite the absence of a line item for Tribal facilities above. EPA, recognizing that the mission of Tribal facilities may differ to some extent from the mission of State and Federally operated facilities, maintains that operating budgets, standardized for production level, are likely to be similar to those presented in Table IX-3 (approximately 3% and 9% respectively).

New Commercial Facilities

EPA estimated that about 4 percent of regulated facilities do not incur any costs under the final regulation and about 76 percent of facilities incur no land or capital costs. The incremental land and capital costs, where they were incurred, represented less than 0.2 percent of total assets. Based on these results, EPA concludes that this final regulation should not present barriers to entry for new businesses. Section 5.2 of this report provides more information.

Small Businesses

The Small Business Administration (SBA) size standard for aquaculture facilities is $0.75 million per year. Accordingly, a "small business" in the aquaculture sector refers to an operation that generates less than $0.75 million in annual revenues.[2]

For this final regulation, EPA identified 37 facilities belonging to a small businesses and 1 facility belonging to a small nonprofit organization. For the purposes of the RFA, Federal, State, and Tribal governments are not considered small governmental jurisdictions, as documented in the rulemaking record (USEPA, 1999). Thus, facilities owned by these governments are not considered small entities, regardless of their production levels. EPA identified no public facilities owned by small local governments in the analysis.

EPA's economic analysis shows that the final rule will have no adverse economic impacts on commercial facilities, including small businesses. The results of EPA's economic analysis (presented in Section 5.1 of this report) covers all regulated facilities, including both small business and businesses that do not meet SBA's small business definition. EPA estimates there are no impacts as measured by EPA's facility and company closure analysis. EPA projects that no facilities belonging to small businesses will close as a result of this final rule. However, EPA does projects some moderate impacts to facilities owned by small businesses. Four facilities have costs-to-sales ratios in excess of five percent but no facilities have costs-to-sales ratios above 10 percent. All of these 4 facilities use a flow through production system. One small business fails the credit test but does not show a change in financial health.

Given the results of the economic analysis of the effects on small businesses, EPA has certified that this action will not have a significant economic impact on a substantial number of small entities. EPA also conducted outreach to small entities and convened a Small Business Advocacy Review Panel to obtain the advice and recommendations of representatives of the small entities that potentially would be subject to the rule's requirements. Section 6 of this report provides more detailed information.

---

[2] SBA defines a "small business" in the agricultural sectors in terms of average annual receipts (or gross revenue) over a 3-year period.

ES.4.2   Economic Effects to National Markets

EPA was not able to prepare a market model analysis for this rule for reasons described in Section 3.6 of this report. Because EPA was not able to prepare a market model analysis for this rule, the Agency is not able to report quantitative estimates of changes in overall supply and demand for aquaculture products and changes in market prices, as well as changes in traded volumes including imports and exports. Despite this limitation, however, EPA does not expect significant market impacts as a result of this final regulation. EPA's analysis shows that no commercial facilities are projected to close. These estimated impacts coupled with the overall cost of the rule, as compared to the total value of the U.S. aquaculture industry, lead EPA to believe that the effects of this regulation on U.S. aquaculture markets will be modest. Finally, EPA believes that long-term shifts in supply associated with this rule are unlikely given expected continued competition from domestic wild harvesters and low-cost foreign suppliers. Three percent of all noncommercial facilities might experience adverse financial effects associated with the rule.

Foreign trade impacts are difficult to predict, since agricultural exports are determined by economic conditions in foreign markets and changes in the international exchange rate for the U.S. dollar. As discussed in Section 3.6 of this report, the U.S. accounts for about 1 percent of world production by weight. Due to the relatively small market share of U.S. aquaculture producers in world markets, EPA believes that long-term shifts in supply associated with this rule are unlikely given expected continued competition from domestic wild harvesters and foreign suppliers. EPA concludes that the impact of this final rule on U.S. aquaculture trade will not be significant.

The communities where aquaculture facilities are located may be affected by the final regulation if facilities cut back operations; local employment and income may fall, sending ripple effects throughout the local community. As EPA's analysis of this final regulation projects no commercial facility closures as a result of this rule, this indicates that the final rule will have no measurable impact on (1) direct losses in commercial production, revenue, or employment; and (2) local economies and employment rates. Therefore, EPA concludes there will be no measurable local or national impacts in the commercial sector. Should some facilities cut back operations as a result of this final regulation, EPA cannot project how great these impacts would be as it cannot identify the communities where impacts might occur. Even under a worst-case scenario that assumes the total costs of the rule are absorbed by the domestic market, EPA estimates that U.S. aquaculture prices would rise by little more than 1 cent per pound. (Section 5.3 of this report provides more detailed information.) Therefore, EPA does not expect significant market impacts as a result of this final rule.

ES.5   COST-BENEFIT ANALYSIS

Table ES-4 shows the economic value of the environmental benefits EPA is able to monetize (i.e., evaluate in dollar terms). EPA estimates the monetized benefits range from $66,214 to $98,616 per year. Monetized benefit categories are primarily in the areas of improved surface water quality (measured in terms of enhanced recreational value). EPA also identified a number of benefits categories that could not be monetized, including reductions in feed contaminants and spilled drugs and chemicals released to the environment, as well as better reporting of drug usage to permitting authorities. These benefits are described in more detail in Sections 7 and 8 of this report and other supporting documentation provided in the rulemaking record.

These estimated benefits compare to EPA's estimate of the total social costs of the final regulations of about $1.4 million per year. These costs include compliance costs to all regulated facilities and administrative costs to Federal and State governments. EPA estimates the administrative cost to Federal and State governments to implement this rule is about $4,392 per year (USEPA, 2002a, 57909). There may be additional social costs that have not been monetized. The benefit estimates are also expressed as pre-tax 2003 dollars. See Section 4.3 of this report for more information.

Table ES-4
Estimated Pre-Tax Annualized Compliance Costs and Monetized Benefits

Production System	Pre-tax Annualized Cost (Thousands, 2003 dollars)
Social Cost	
Flow Through and Recirculating	$1,406
Net Pen	$36
Subtotal (Industry Costs)	**$1,442**
State and Federal Permitting Authorities	$3
Estimated Total Costs	**$1,445**
Monetized Benefits	
	$66 to $99
Estimated Total Benefits	**$66 to $99**

Note: Totals may not sum due to rounding
*Monetized benefits are not scaled to the national level.

ES.6   REFERENCES

R.A. Brealy and S.C. Myers. 1996. *Principles of Corporate Finance.* 5th edition. The McGraw-Hill Companies, Inc. New York.

Brigham, E.F., and L.C. Gapenski. 1997. *Financial Management: Theory and Practice.* 8th edition. The Dryden Press. Fort Worth, Texas.

ENR (Engineering News Record). 2004. Construction Cost Index History (1908-2004). <http://www.enr.com/cost/costcci.asp>.

FFSC (Farm Financial Standards Council). 1997. *Financial Standards for Agricultural Producers.* December. DCN 20095.

Jarnagin, Bill D. 1996 *Financial Accounting Standards: Explanation and Analysis.* 18th edition. CCH, Incorporated. Chicago, IL.

USEPA (United States Environmental Protection Agency). 2004. *Technical Development Document for the Final Effluent Limitations Guidelines and Standards for the Aquatic Animal Production Industry.* Washington, DC: U.S. Environmental Protection Agency, Office of Water. [EPA-821-R-04-012]

USEPA (U.S. Environmental Protection Agency). 2003. Effluent Limitations Guidelines and New Source Performance Standards for the Concentrated Aquatic Animal Production Point Source Category; Notice of Data Availability; Proposed Rule. 40 CFR Part 451. *Federal Register* 68:75068-75105. December 29.

USEPA (U.S. Environmental Protection Agency). 2002a. Effluent Limitations Guidelines and New Source Performance Standards for the Concentrated Aquatic Animal Production Point Source Category; Proposed Rule. 40 CFR Part 451. *Federal Register* 67:57872. September 12.

USEPA (U.S. Environmental Protection Agency). 2002b. Economic and Environmental Impact Analysis of the Proposed Effluent Limitations Guidelines and Standards for the Aquatic Animal Production Industry. Washington, DC: U.S. Environmental Protection Agency, Office of Water. EPA-821-R-02-015. September.

USEPA (United States Environmental Protection Agency). 2002c. Development Document for the Proposed Effluent Guideline and Standards for the Aquatic Animal Production Industry. EPA-821-R-02-016. Washington, DC: U.S. Environmental Protection Agency, Offices of Water.

USEPA (United States Environmental Protection Agency). 2002e. Detailed Questionnaire for the Aquatic Animal Production Industry. Washington, DC: OMB Control No. 2040-0240. Expiration Date November 30, 2004.

USEPA (United States Environmental Protection Agency). 2001. Screen Questionnaire for the Aquatic Animal Production Industry. OMB Control No. 2040-0237. Washington, DC. July.

USEPA (United States Environmental Protection Agency). 2000. *Guidelines for Preparing Economic Analyses.* Washington, DC: U.S. Environmental Protection Agency. EPA 240-R-00-003. September. DCN 20435.

USEPA (U.S. Environmental Protection Agency). 1999. *Revised Interim Guidance for EPA Rulewriters: Regulatory Flexibility Act as amended by the Small Business Regulatory Enforcement Fairness Act.* Washington, DC: U.S. Environmental Protection Agency. 29 March. EPA Docket No. OW-2002-026, DCN 20121.

CHAPTER 1

INTRODUCTION

1.1   SCOPE AND PURPOSE

The U.S. Environmental Protection Agency (EPA) proposes and promulgates water effluent discharge limits (effluent limitations guidelines and standards) for industrial sectors. This document summarizes both the costs, economic impacts, and benefits of technologies that form the bases for the final limits and standards for the concentrated aquatic animal production (CAAP) industry.

The Federal Water Pollution Control Act (commonly known as the Clean Water Act [CWA, 33 U.S.C. §1251 et seq.]) establishes a comprehensive program to "restore and maintain the chemical, physical, and biological integrity of the Nation's waters" (section 101(a)). EPA is authorized under sections 301, 304, 306, and 307 of the CWA to establish effluent limitations guidelines and standards of performance for industrial dischargers. The standards EPA establishes include:

- Best Practicable Control Technology Currently Available (BPT). Required under section 304(b)(1), these rules apply to existing industrial direct dischargers. BPT limitations are generally based on the average of the best existing performances by plants of various sizes, ages, and unit processes within a point source category or subcategory.

- Best Available Technology Economically Achievable (BAT). Required under section 304(b)(2), these rules control the discharge of toxic and nonconventional pollutants and apply to existing industrial direct dischargers.

- Best Conventional Pollutant Control Technology (BCT). Required under section 304(b)(4), these rules control the discharge of conventional pollutants from existing industrial direct dischargers.[1]   BCT replaces BAT for control of conventional pollutants.

- Pretreatment Standards for Existing Sources (PSES). Required under section 307(b). Analogous to BAT controls, these rules apply to existing indirect dischargers (whose discharges flow to publicly owned treatment works [POTWs]).

- New Source Performance Standards (NSPS). Required under section 306(b), these rules control the discharge of toxic and nonconventional pollutants and apply to new source industrial direct dischargers.

- Pretreatment Standards for New Sources (PSNS). Required under section 307(c). Analogous to NSPS controls, these rules apply to new source indirect dischargers (whose discharges flow to POTWs).

---

[1] Conventional pollutants include biochemical oxygen demand (BOD), total suspended solids (TSS), fecal coliform, pH, and oil and grease.

Prior to this rule, EPA defined "concentrated aquatic animal production facilities" at 40 CFR 122, Appendix C, and identified the need for them to obtain National Pollutant Discharge Elimination System (NPDES) permits, but had not set national effluent limitations guidelines or standards for these or a subset of these dischargers.

1.2 DATA SOURCES FOR THE FINAL RULE

EPA's economic analysis relied on a wide variety of data and information sources. Data sources used in the economic analysis include:

- ■ EPA's Screener Questionnaire for the Aquatic Animal Production Industry (USEPA, 2001)

- ■ EPA's Detailed Questionnaire for the Aquatic Animal Production Industry (USEPA, 2002)

- ■ U.S. Department of Agriculture (USDA), particularly USDA's *1998 Census of Aquaculture* (USDA, 2000)

- ■ Joint Subcommittee on Aquaculture (JSA). JSA is a Federal interagency coordinating group to increase the overall effectiveness and productivity of Federal aquaculture research, technology transfer, and assistance programs. It was authorized under the National Aquaculture Act of 1980 and the National Aquaculture Improvement Act of 1985. (For more information see: http://ag.ansc.purdue.edu/aquanic/jsa/).

- ■ Academic literature

- ■ Industry journals

- ■ General economic and financial references

The use of each of these major data sources is discussed below.

EPA collected facility-level production data from individual aquatic animal producers through a screener survey administered under the authority of the CWA Section 308 (USEPA, 2001). EPA used response data from the screener survey to classify and subcategorize facilities by production method, species produced and production level, and water treatment practices in place prior to the proposed regulation. EPA identified the subset of concentrated aquatic animal production facilities deemed to be in scope of the proposed rule.

EPA used the information from the screener survey to identify a subset of facilities to receive the detailed questionnaire. Like the screener survey, EPA administered the detailed survey under the authority of the CWA Section 308 (USEPA, 2002). EPA used response data from the survey to classify and subcategorize facilities by production method, species produced and production level, and water treatment practices in place prior to the proposed regulation. For commercial operations, the survey instrument collected financial and economic information at the aquaculture enterprise, the facility, and the company that owned the facility. For public or noncommercial operations, EPA collected financial and economic information on operating costs and funding sources. Due to the timing of the surveys and the rulemaking schedule, the proposal analysis was based on the screener survey data while the detailed survey formed the basis for the results presented in the Notice of Data Availability (USEPA, 2003) and for final promulgation.

EPA relied heavily on the USDA *1998 Census of Aquaculture* to profile the industry at proposal (USDA, 2000). EPA relied on the Census for the national number of aquaculture facilities, which establishes a starting point to evaluate EPA's regulatory flexibility.

The Joint Subcommittee on Aquaculture (JSA) formed an Aquaculture Effluents Task Force (AETF) to assist EPA. The Economics Subgroup provided enterprise budgets, additional references, industry literature and journal articles to EPA. An enterprise budget depicts financial conditions for representative aquaculture facilities. Enterprise budgets are useful tools for examining the potential profitability of an enterprise prior to actually making an investment. To create an enterprise budget, an analyst gathers information on capital investments, variable costs (such as labor and feed), fixed costs (e.g., interest and insurance), and typical yields and combines it with price information to estimate annual revenues, costs and return for a project. By varying different input parameters, enterprise budgets can be used to examine the relative importance of individual parameters to the financial return of the project or to identify breakeven prices required to provide a positive return. The Economics Subgroup provided EPA with enterprise budgets or reports for trout, shrimp, hard clams, prawns, and alligators (Docket OW-2002-0026, Section 8.2.3 DCNs 20073, 20080, 20082, 20084, 20131, and 20132).

EPA used academic journals and industry sources such as trade journals and trade associations to develop its industry profile to formulate a better understanding of industry changes, trends, and concerns. As necessary, EPA cites various economic and financial references used in its analysis throughout this report. These references may be in the form of financial and economic texts, or other relevant sources of information germane to the impact analysis.

1.3   OVERVIEW OF CHANGES TO EPA'S ECONOMIC METHODOLOGY

For the proposed rule, EPA evaluated projected economic impacts using screener questionnaire data which did not include financial or economic information beyond revenues and limited production data. As a consequence, the proposal's impact analysis was based on compliance costs for model facilities, frequency factors for extrapolating costs to a group of facilities represented by a model, and sales or revenue tests. Revenue tests involve simple comparisons of compliance costs with facility revenues. For noncommercial facilities, in lieu of revenues, EPA imputed a value to their production based on annual harvest and commercial prices. Similar revenues tests were applied to both commercial and noncommercial facilities. EPA estimated the number of small businesses from a special tabulation of USDA's *1998 Census of Aquaculture* (USDA/NASS, 2002).

For the final rule EPA is able to conduct a more detailed financial impact analysis because of the availability of facility-specific pairs of costs and revenues collected in the detailed questionnaire after proposal. The availability of these data permit a more detailed analysis for different subpopulations within the regulated community within the scope of this rule, including both commercial and noncommercial aquaculture facilities.

1.4   REPORT ORGANIZATION

This report is organized as follows:

- Chapter 2—EPA Detailed Questionnaire. Summarizes information EPA collected in the detailed questionnaire for the facilities considered within the scope of the final rule.

- Chapter 3—Economic Methodology. Summarizes EPA's methodology to examine incremental pollution control costs and their associated economic impacts.

- Chapter 4—Regulatory Options: Descriptions, Costs, and Conventional Pollutant Removals. Presents a brief description of the regulatory options considered by EPA. More detail is given in the *Development Document* (USEPA, 2004).

- Chapter 5—Economic Impact Results. Presents the results of EPA's analysis of the estimated annual costs and the economic impacts on regulated facilities associated with the final regulations, using the methodology presented in Chapter 3.

- Chapter 6—Small Entity Flexibility Analysis. Presents the results of EPA's analysis of the possible financial effects on small businesses that are affected by the final regulations, as required under the Regulatory Flexibility Act as amended by the Small Business Regulatory Enforcement Fairness Act

- Chapter 7—Environmental Assessment. Briefly describes effluent quality and loads from CAAP facilities, and summarizes literature relating to water quality and aquatic ecosystem effects of aquaculture effluents.

- Chapter 8—Environmental Benefits of Final Regulation. Summarizes the methods and results for estimating monetized benefits associated with the rule.

- Chapter 9— Other Regulatory Analysis Requirements. Presents EPA's assessment of the nationwide costs and benefits of the regulation pursuant to Executive Order 12866 and the Unfunded Mandates Reform Act (UMRA).

1.5   REFERENCES

USDA (U.S. Department of Agriculture, National Agricultural Statistics Service). 2000. *1998 Census of Aquaculture*. Also cited as 1997 Census of Agriculture. Volume 3, Special Studies, Part 3. AC97-SP-3. February.

USDA/NASS (U.S. Department of Agriculture, National Agricultural Statistics Service). 2002. Special tabulation request submitted to USDA NASS. Information relayed to EPA and Eastern Research Group, Inc. March 6.

USEPA (U.S. Environmental Protection Agency). 2004. *Development Document for the Final Effluent Limitations Guidelines and Standards for the Aquatic Animal Production Industry.* Washington, DC: U.S. Environmental Protection Agency, Office of Water.

USEPA (U.S. Environmental Protection Agency). 2003. Effluent Limitations Guidelines and New Source Performance Standards for the Concentrated Aquatic Animal Production Point Source Category; Notice of Data Availability; Proposed Rule. 40 CFR Part 451. *Federal Register* 68:75068-75105. December 29.

USEPA (U.S. Environmental Protection Agency). 2002. Detailed Questionnaire for the Aquatic Animal Production Industry. Washington, DC: OMB Control No. 2040-0240. Expiration Date November 30, 2004.

USEPA (U.S. Environmental Protection Agency). 2001. Screener Questionnaire for the Aquatic Animal Production Industry. Washington, DC: OMB Control No. 2040-0237. Expiration Date July 26, 2004.

CHAPTER 2

EPA DETAILED QUESTIONNAIRE SURVEY

In August 2001, EPA mailed a short screener survey "Screener Questionnaire for the Aquatic Animal Production Industry" to approximately 6,000 aquatic animal production facilities (USEPA, 2001). EPA received responses from 4,900 facilities with about 2,300 facilities reporting that they produce aquatic animals. EPA used the screener survey information to select a stratified random sample of this industry to receive a detailed questionnaire (USEPA, 2002a). The sample included pond systems, aquariums, trout or salmon production for facilities that produce more than 20,000 pounds per year (lbs/yr). The sample also included other facilities that are not in scope of the rule, given information in the 2001 survey. EPA included such facilities to re-examine its proposal on the scope of the regulation. EPA describes the criteria for the inclusion in the sample frame along with the number of questionnaires mailed out, returned, and usable in the December 29, 2003 Notice of Data Availability ("Notice") (USEPA, 2003; FR 68:75072). The Notice presents the facility counts, costs, and impacts for net pen, flow-through, and recirculating systems that produce more than 20,000 lbs/yr of trout or salmon or more than 100,000 lbs/yr of other biomass (USEPA, 2003; FR 68:75093-75100).

The data presented in this Chapter represent those facilities that EPA determined to be in the scope of the final rule. That is, the facilities meet two criteria: (1) use net pen, flow-through, or recirculating systems, and (2) produce more than 100,000 lbs/yr. Section 2.1 summarizes the estimated facility counts and how the change from screener survey data to detailed survey data changed the profile of in-scope facilities. The facility counts also highlight the relative roles of the commercial and non-commercial sectors in the aquaculture industry. Section 2.2 contains the information for commercial facilities; Section 2.3 reports the data for noncommercial facilities.

The information in Sections 2.2 and 2.3 is presented separately for flow-through and recirculating systems because the financial characteristics for these two sets of observations differ slightly. For technical reasons, however, the industry is subcategorized into "continuous discharge" (i.e., flow-through and recirculating systems) and "net pens." For a more complete discussion of the aquaculture industry as a whole, see the industry profile in the proposal EEIA (USEPA, 2002a, Chapter 2).

2.1   FACILITY COUNTS

The U.S. Department of Agriculture reported that there were about 4,000 commercial aquaculture facilities nationwide in 1998 (USDA, 2000). EPA estimates that the number of non-commercial facilities is between 530 to 690 Federal, State, Tribal, and Academic/Research facilities (USEPA, 2002b, see Table 2-2).[1]

Not all aquaculture facilities are affected by the final regulation. EPA estimates that there are approximately 242 "in-scope" facilities that will be affected by the rule. Regulated facilities are

---

[1] EPA estimates that there are about 320 noncommercial facilities with net pen, recirculating, or flow-through systems that produce more than 20,000 lbs/yr. (USEPA, 2003; 68 FR 75093).

comprised of 101 commercial (less than 1 percent of all commercial facilities nationwide) and 141 noncommercial (between roughly one-third and one-fourth of all noncommercial facilities). In terms of annual production, EPA estimates that this final regulation affects about 17 percent of total aquacultural production in the United States.[2]

Table 2-1 and Figure 2-1 summarize the national number of facilities estimated to be within the scope of the final rule. The number of facilities shown for commercial and noncommercial in Table 2-1 are estimated with the sample weights derived from the detailed questionnaire. That is, EPA sent detailed questionnaires to some but not all of the facilities withing a stratified sampling plan. Facilities within a stratum share common characteristics so data collected from some of them can be used to extrapolate to all facilities within the stratum by using the facility weights.

The noncommercial group includes Federal, State, and Tribal facilities. No Tribal facilities that returned a detailed questionnaire produced over the 100,000 lb/yr threshold but, if such a facility exists among the facilities that did not receive a detailed questionnaire, it would likely resemble other noncommercial facilities within the scope of the rule.

Facilities in Alaska are different. They practice ocean ranching rather than aquaculture and, although they are not for profit organizations, they report revenue from harvested salmon that return to the release area. EPA identified two Alaska nonprofit facilities that are within the scope of the rule but which were not selected to receive a detailed questionnaire. These are listed separately in Table 2-1. Because they are not represented in the detailed survey, they are not included in the discussion of noncommercial facilities in Section 2.2.

EPA identified no academic or research facilities within the scope of the final rule.

---

[2] Based on an estimated 94 million pounds of production by commercial facilities and an estimated 43.3 million pounds of production by noncommercial facilities, as compared to a total U.S. production of estimated total U.S. aquaculture production of about 820 million pounds in 2001 (NMFS, 2003).

Table 2-1
Estimated Number of In-Scope Facilities by Organization and Production System

Organization	Production System	Estimated Number of Facilities
Commercial[1]	Flow-Through	70
	Recirculating	12
	Net Pen	19
Noncommercial[1]	Flow-Through	138
	Recirculating	1
	Net Pen	0
Alaska Nonprofit	Flow-Through	2
TOTAL		242

[1]EPA estimates from detailed survey (USEPA, 2002a).

With the additional information collected in the detailed survey, EPA decided to restrict the scope of the rule to facilities using flow-through, recirculating, or net pen systems that produce more than 100,000 lbs/yr. See the *Development Document* for more details (USEPA, 2004).

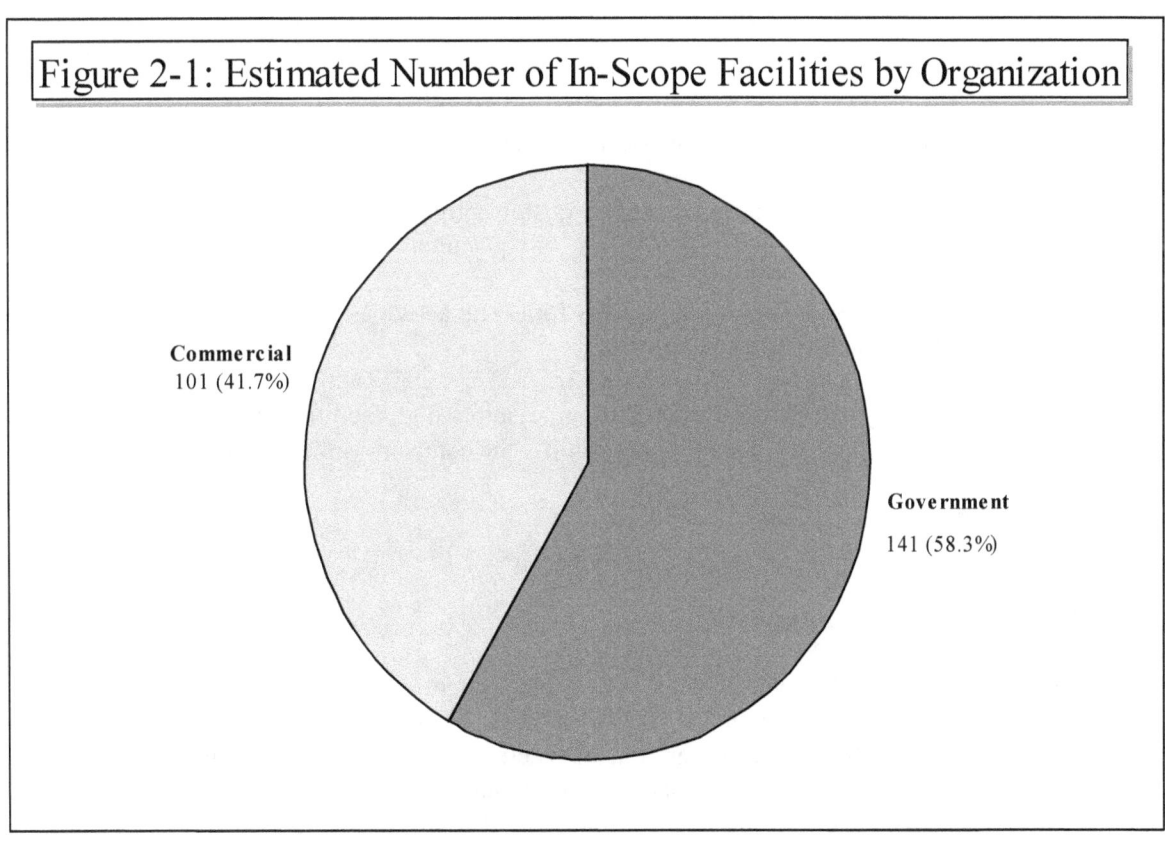

Figure 2-1: Estimated Number of In-Scope Facilities by Organization

Commercial
101 (41.7%)

Government
141 (58.3%)

Table 2-2
In-Scope Commercial Facilities by Geographic Distribution

State	Number of In-Scope Facilities
California	5
Colorado	6
Georgia	4
Idaho	25
Maine	22
Montana	4
North Carolina	13
Pennsylvania	5
Texas	3
Virginia	4
Washington	4
Other States	6
Total	101

Source: EPA estimates from detailed survey (USEPA, 2002b).

2.2   COMMERCIAL FACILITIES

EPA collected information at various organizational levels for commercial facilities:

■   Enterprise: Responses reflect fish-raising activities only where the facility or farm is engaged in other businesses outside of aquaculture at the same site.

■   Facility: Responses for the facility reflect all activities. If the only activity is raising fish, the facility is the enterprise.

■   Company: Responses reflect the aggregation of all facilities under the same ownership. If there is only one facility, the company is the facility.

Data for each of these levels are summarized below.

2.2.1   Enterprise Information

EPA found eight facilities within the scope of this effluent guideline that reported other business activities. This represents about 8 percent of all commercial facilities affected by the rule. For all of these facilities, aquaculture is the primary business activity; however some facilities also engaged in other agricultural activities such as raising livestock. For about half of the facilities, aquaculture was the profitable enterprise, while the other enterprises lost money. These eight facilities employ 80 people and generated revenues of $6.3 million in 2001. Due to the small number of facilities, no further detail is provided for reasons of confidentiality.

2.2.2   Facility Information

Nationally, EPA identified 101 commercial facilities within the scope of this regulation. These facilities operate in many different regions and raise a wide variety of species for a number of different markets. Some of these markets include food size fish for sale to wholesalers, restaurants, retailers, or private game and sport clubs for stocking ponds, or eggs, fry, and fingerlings for sale to other aquaculture facilities. The EPA survey did not collect data regarding the "point of first sale," e.g., a processor or a restaurant, but it is likely that some producers sell their product to more than one type of customer.

2.2.2.1   Geographic Distribution

Commercial facilities in the detailed questionnaire database and within scope of this regulation are spread over 16 states. Idaho, Maine, and North Carolina have the largest number of in-scope facilities at 25, 22, and 13 facilities, respectively. Table 2-2 summarizes the number of facilities by state for in-scope commercial facilities.

2.2.2.2   Revenues

EPA's detailed survey collected financial data for a 3-year period, 1999-2001. EPA estimates that commercial facilities within the scope of EPA's final rule generated over $155 million in 2001. NMFS estimated 2001 total U.S. aquaculture production as about $935 million (NMFS, 2003).[3] Revenues reflect a variety of sales end-points, e.g., sales to processors, direct sales to markets and restaurants, and stocker sales to fee fishing operators.

EPA estimates that net pen facilities have the largest total revenue ($89 million), average revenue ($4.4 million per facility), and generate 57 percent of the total industry revenues. Recirculating facilities generated an estimated $25.8 million in 2001 (17 percent of the industry total). The average recirculating facility generated $2.1 million in 2001. Flow-through facilities accounted for an estimated 26 percent of total revenue ($40.3 million) in 2001; however the average flow-through facility generated significantly smaller revenues compared to net pen and recirculating facilities. The average net pen facility revenue was more than 7 times larger than the average flow-through, while the average recirculating was almost 3.5 times larger than the average flow though. Table 2-3 summarizes the total and average revenues for commercial facilities by production system.

---

[3]Thus EPA set the scope of the rule to affect less than 3 percent of the commercial facilities yet regulate about 17 percent of the value of production.

Table 2-3
Estimated 2001 Revenues for Commercial In-Scope Facilities by Production System

Production System	Number of In-Scope Facilities	Estimated Revenues (Millions, 2001 dollars)	Estimated Facility Revenue* (Millions, 2001 dollars)
Flow-through	70	$40.3	$0.6
Recirculation	12	$25.8	$2.1
Net pen	19	$89.0	$4.4
Total	101	$155.1	$1.5

Source: EPA estimates from detailed survey (USEPA, 2002b).
*Average facility revenue is not an average of the total numbers presented in this table. Some facilities either did not provide revenue data, or could not separate facility data from company revenue information. To represent the average accurately EPA took the average based on only those facilities able to provide information.

Over the three year period (1999-2002) for which EPA collected financial data, flow-through and recirculating facilities experienced growth in revenues (Table 2-4). Net pens, in contrast, experienced a increase in revenues from 1999 to 2000 followed by a decrease in revenues in 2001. This is likely due to a combination of an outbreak of infectious salmon anemia (ISA) requiring the destruction of many stocks and lower prices for the stocks surviving in other regions.

Table 2-4
Estimated Revenues for Commercial In-Scope Facilities by Production System, 1999-2001

Production System	Number of In-Scope Facilities	Estimated Total Revenues (Millions)		
		2001	2000	1999
Flow-through	70	$40.3	$39.0	$34.3
Recirculating	12	$25.8	$24.7	$21.4
Net pen	19	$89.0	$125.1	$111.4
Total	101	$155.1	$188.8	$167.1

Source: EPA estimates from detailed survey (USEPA, 2002b).
Note not all facilities were able to provide data for all three years. Numbers presented here reflect data facilities were able to report.

2.2.2.3 Production

EPA's detailed survey collected production data for three years (1999, 2000, and 2001). The data reflected the life cycle stage (egg, fry, fingerling, stocker, foodsize, or broodstock) and measurement unit used by the respondent (e.g., count or pounds). EPA converted these responses to pounds using the conversion factors presented in Table 2-5. For more information see the *Development Document* supporting the proposed rulemaking (USEPA, 2002c). To convert counts to pounds, EPA multiplied production counts by the conversion factor for the specified species and size.

Table 2-5
Conversion Factors for Reporting Production in Pounds (Abridged)

Species	Size					
	Foodsize	Stockers	Fingerlings	Seed Stock	Brood-Stock	Fry
Catfish	1.50	0.18	0.0334	0.0001	4.3100	0.0014
Trout	1.00	0.32	0.0350	0.0001	2.5000	0.0014
Salmon	5.00	0.32	0.0350	0.0001	10.0000	0.0014
Striped Bass	1.75	0.32	0.0600	0.0001	5.0000	0.0014
Tilapia	1.75	0.32	0.0350	0.0001	2.5000	0.0014
Bass	2.00	0.1418	0.0214	0.0001	3.4247	0.0014
Sturgeon	45.00	0.1418	0.0214	0.0001	3.4247	0.0014
Sunfish	0.25	0.1418	0.0214	0.0001	3.4247	0.0014
Walleye	3.00	0.1418	0.0214	0.0001	3.4247	0.0014
Pike	4.63	0.1418	0.0214	0.0001	3.4247	0.0014
Carp (includes koi, white amour)	4.00	0.1418	0.0214	0.0001	3.4247	0.0014
Shrimp	0.0444	0.1418	0.0214	6.6E-6	0.1000	0.0014

Source: USEPA, 2002c.

Table 2-6 presents EPA's estimate of 2001 production by production system. In that year, EPA estimates that commercial facilities within the scope of this effluent guideline produced about 94 million pounds of fish and other aquatic animals.[4] Over 74 percent of that production was raised in net pen systems. Net pen facilities also have the highest average production with more than three million pounds per year. Their average production is more than four times higher than the average recirculating facility production and more than 12 times greater than those for the average flow-through facility.

---

[4] The in-scope population contains at least one example of crustaceans raised in a flow-through, recirculating, or net pen system.

Table 2-6
Estimated 2001 Production Data for In-Scope Commercial Facilities by Production System

Production System	Number of In-Scope Facilities	Estimated Production (Million lbs)	Estimated Average Facility Production (Million lbs/facility)*
Flow-Through	70	14.7	0.2
Recirculating	12	9.2	0.7
Net pen	19	70.0	3.4
Total	101	93.9	0.8

Source: EPA estimates from detailed survey (USEPA, 2002b).
*Average facility production is not an average of the total numbers presented in this table. Some facilities either did not provide production, or could not separate facility from company production. To represent the average accurately, EPA calculated the average based only on those facilities able to provide information.

Table 2-7 presents EPA's estimate of aggregate production by production system for 1999-2001. Some facilities did not report production, therefore the production numbers presented here may be underestimated. Trends varied by production system. When a flow-through facility reported production, the facility reported three years of data. Between 1999 and 2001, flow-though facilities decreased production by roughly 2 million pounds (Table 2-7), although estimated revenues increased over the same time period (Table 2-4). This implies flow-through facilities may have been able to charge higher prices in 2001 than 1999 or shifted sales to a more profitable outlet (e.g., from processors to restaurants). Production by recirculating facilities increased by about one million pounds while revenues also increased.[5] Between 19999 and 2001, net pen production, as a whole, increased, but this is attributable to several facilities reporting production for the first time (e.g., start-up facilities or a change in ownership). Revenues among net pen operations decreased over this period (compare with Table 2-4), implying lower prices.

Table 2-7
Estimated Aggregate Production Data for In-Scope Commercial Facilities by Production System

Production System	Number of In-Scope Facilities	Estimated Production (Million lbs)		
		2001	2000	1999
Flow-Through	70	14.7	14.9	16.2
Recirculation	12	9.2	9.0	8.6
Net pen*	19	70.0	68.6	56.7
Total	101	93.9	92.5	81.5

Source: EPA estimates from detailed survey (USEPA, 2002b).
NOTE: Not all facilities reported production in the survey, therefore numbers presented here are likely to be underestimates.

---

[5]Most of the increase in production and revenue is attributable to one facility that did not report results for 1999. All other recirculating facilities reported three years of data.

2.2.2.4 Employment (Paid and Unpaid)

EPA collected data on 2001 employment, full and part-time, paid and unpaid, for all commercial facilities in the detailed survey. Numbers presented are in full time equivalents (FTE). That is, two part-time employees working 20 hours a week are presented as one full time employee.

Nationally, EPA projects there to be nearly 955 people employed at commercial facilities within the scope of the final regulation (Table 2-8). Total flow-through employment in 415 FTEs, indicating that flow-though facilities provide 43 percent of the jobs within the scope of the regulation. However, this is due to the large number of flow-through facilities not because of reported high employment at each facility. On average, a flow-through facilities employ about 6 people per facility. In contrast, recirculating facilities employ the smallest total number (reported at 172 FTEs), but the average facility employs about 14 people, more than twice as many as an average flow-through facility. The average net pen facility employs the most people at about 23 per facility with a total employment of 371 FTEs.

EPA also asked each survey respondent to differentiate unpaid labor from paid labor and/or management. Of the 101 commercial facilities within the scope of this final regulation, only 3 facilities reported unpaid labor and/or management (Table 2-8).[6] All three facilities are flow-through facilities. For the purpose of completing Table 2-8, EPA assumes at least one employee per facility. See Appendix A for a more detailed discussion of unpaid labor and/or management in the economic analysis.

2.2.2.5 Costs and Returns for Flow Through and Recirculating Commercial Facilities

Table 2-9 presents a summary of the national estimates of the 2001 costs and returns for commercial facilities in the flow through and recirculating subcategory that produce more than 100,000 pounds of aquatic animals per year. These data are based on costs and returns information from EPA's detailed questionnaire (USEPA, 2002b), and include both operations considered baseline failures and facilities that remain profitable in the baseline analysis. Among these facilities, there are 82 in-scope facilities with flow-through or recirculating production systems in the EPA detailed questionnaire data base. Four facilities cannot be analyzed for costs and returns because an unweighted facility changed ownership at the time of the survey and did not supply financial data. The number of observations for this analysis is 78. Although 3 years of financial information were collected by EPA, the data in the table are for 2001 since this is the most recent year for which data were collected.

The average sales estimate for operations that produce between 100,000 lbs/yr and 475,000 lbs/yr is somewhat skewed because two unweighted facilities reported $0 sales in their survey. This is attributable to the fact that production from these two facilities are transferred to other facilities under the same ownership for further grow-out and or processing. Also, the survey data indicate that not all cost components are individually tracked by the survey respondents, as evidenced by the frequent minimum value of $0 for several cost categories.

---

[6]EPA's decision to define the scope of the analysis to facilities producing more than 100,000 lbs/yr had the effect of removing 44 additional facilities that reported unpaid labor and/or management from the potentially affected population. For the remaining three facilities, EPA examined the effects of including three different estimates for labor costs (e.g., Federal minimum wage, Bureau of Labor Statistics wages for farm managers, and USDA's Agricultural Resource Management Survey (ARMS) estimates for commercial farms) on the economic analysis for those facilities. None of imputed labor costs affected the impacts estimated as a result of the rule (ERG, 2004).

Table 2-8
Total and Average Employment for In-Scope Commercial Facilities by Production System

Production System	Number of In-Scope Facilities	Employment (FTE)		Reported Average Facility Employment (FTE)
		Paid	Unpaid	
Flow-Through	70	412	3	6
Recirculation	12	172	0	14
Net pen	19	371	0	23
Total	101	955	3	10

Source: EPA estimates from detailed survey (USEPA, 2002b).

Table 2-9
National Estimates of Costs and Returns at In-scope, Flow Through or Recirculating, Commercial Facilities, 2001

Variable	>475,000 lbs/yr (30 facilities)			100,000 lb/y - 475,000 lbs/yr (59 facilities)		
	National Average	Minimum	Maximum	National Average	Minimum	Maximum
Total Sales	$1,456,563	$358,000	$9,766,000	$441,863	$140,000	$1,456,000
Total Expenses	$1,619,269	$381,000	$10,573,000	$564,332	$138,000	$1,610,000
All Variable	$1,048,959	$304,000	$4,311,000	$353,669	$69,000	$1,232,000
Depreciation	$92,510	$3,000	$669,000	$52,095	$0	$304,000
Feed	$201,696	$0	$422,000	$88,928	$0	$249,000
Chemicals	$65,600	$0	$377,000	$13,110	$0	$146,000
Non-Mortgage Interest	$63,026	$0	$612,000	$5,097	$0	$44,000
Labor	$243,623	$0	$1,925,000	$91,718	$0	$503,000
Rent-Vehicle	$3,312	$0	$17,000	$2,722	$0	$29,000
Rent-land	$71,216	$0	$277,000	$12,279	$0	$68,000
Repairs	$39,498	$0	$227,000	$24,226	$0	$103,000
Energy	$119,541	$0	$801,000	$54,793	$0	$281,000
COGS	$562,405	$0	$7,861,000	$49,833	Cash basis	$438,000
All Fixed	$46,314	$0	$346,000	$44,275	$0	$161,000
Taxes	$23,869	$0	$230,000	$22,026	$0	$69,000
Interest-Mortgage	$6,219	$0	$18,000	$1,952	$0	$16,000
Insurance	$16,226	$0	$116,000	$20,297	$0	$9,200

Source: USEPA detailed survey (USEPA, 2002b).
Variable Costs: Labor (hired labor, including management if paid); Feed purchased (production and medicated); Chemicals (including fertilizers and lime); Energy (utility costs, gasoline, fuel, and oil); Depreciation; Interest (other than mortgage interest, although may or may not be interest on operating loan); Repairs/maintenance; Cost of aquatic animals (only if the respondent used accrual accounting); Other rent or lease (vehicles, machinery, equipment, land, animals, etc.). Fixed Costs: Insurance (other than health); Interest (mortgage interest); Taxes.

2.2.2.6 "Captive Facilities"

A site is classified as "captive" when a certain percentage of its production is shipped to other sites under the same ownership. EPA found seven such commercial sites that ship all of their production to other sites under the same ownership. All of these sites are salmon hatcheries that exist solely to supply fingerlings to net pen sites under the same ownership for grow-out. For these facilities, the closure analysis defaults to the company level.

2.2.3   Company Information

At the company level, EPA collected information on organization type, revenues (for 2001), and assets (for 2001). EPA collected company-level revenue data because Small Business Administration (SBA) sets the small business standard for this industry as $750,000 in annual revenues with revenues as reported at the top of the corporate hierarchy, not the facility level (i.e., it is possible that a large company is made up of a number of "small" facilities, see SBA, 2001). EPA also collected data on assets and liabilities to use USDA's methodology for evaluating farm financial health using company-level debt/assets ratios.

2.2.3.1 Number of Companies

The 101 in-scope commercial facilities are a national estimate calculated by multiplying the raw data from 37 unweighted in-scope commercial facilities by statistical survey weights. EPA reviewed the 37 unweighted in-scope commercial facilities that received a detailed survey to determine their corporate parent. These facilities were owned by 34 companies. Of these 34 companies, 30 are single-facility companies. The statistical weights, however, are developed on the basis of facility characteristics and therefore cannot be used for estimating the number of companies. Hence, it is not appropriate to combine the two sets of counts (e.g., it is not appropriate to divide 101 facilities by 34 companies to arrive at slightly under 3 facilities per company). The domestic industry is characterized by companies operating only one site.

Most of the multi-site companies raise salmon. These operations consist of either multiple net pen sites or a combination of flow-through hatcheries and net pen grow-out sites (e.g., "captive" facilities).

2.2.3.2 Company Organization

The 34 companies owning the 37 (unweighted) in-scope commercial facilities are organized as:

- ▪ 16 C corporations
- ▪ 12 S or limited liability corporations
- ▪ 2 limited partnerships
- ▪ 4 sole proprietorships

One of the 34 companies is publicly-held. All others are either privately-held or foreign. Public data on foreign firms would not include details on specific U.S.-based operations. EPA's survey is, therefore, the only source of financial information for the U.S. divisions of foreign firms and privately-held companies. The foreign companies are concentrated in the salmon industry.

2.2.3.3 Revenues

EPA is not presenting company-level revenues for reasons of confidentiality. Of the 34 (unweighted) companies, EPA assumes that 30 companies in the economic analysis are single-facility companies. Reporting company revenues might compromise the revenue information for the four multi-facility companies.

2.2.3.4 Number of Small Businesses

Of the 34 companies, 11 report $750,000 or less in annual revenues, i.e., they meet the SBA definition of a small business among aquaculture facilities (SBA, 2001).

2.2.3.5 Assets

EPA collected balance sheet information at the company-level for all companies (Table 2-10). Since most companies operate only one site or utilize only one production system, EPA was able to separate assets by production system used for purposes of comparison. In a few cases, a company owned more than one facility and the company assets represented more than one production system. An example would be flow-through salmon facilities that produce smolts or fingerlings which are then transferred to net pens for grow-out. Because salmon sales from the net pen facilities form most, if not all, company sales, all company assets are allocated to the net pen production systems for the example company.

Companies operating flow-through facilities make up 59 percent of the total companies, but account for only 10 percent of total industry assets. Based on EPA's detailed survey, the average company operating a flow-through facility has $633,000 in assets. In contrast, net pen companies, 24 percent of all companies, account for over 71 percent of total assets ($88.6 million) and also have the largest average assets ($11.1 million). The average net pen facility has more than 18.5 times the assets of a flow-through facility and more than three times the assets of a recirculating facility. Companies operating primarily recirculating constitute 17 percent of all companies, and 18 percent of total assets, with the average recirculating company possessing about $3.7 million dollars in assets.

Table 2-10

Total and Average 2001 Assets Reported by In-Scope Commercial Facilities by Production System

Production System	Number of (Unweighted) In-Scope Companies	2001 Assets (Millions)	2001 Average Assets (Millions)
Flow-Through	20	$12.6	$0.6
Recirculating	6	$22.2	$3.7
Net pen	8	$88.6	$11.1
Total	34	$123.4	$15.4

Source: EPA estimates from detailed survey (USEPA, 2002b).

2.3   NONCOMMERCIAL FACILITIES

Noncommercial facilities raise aquatic animals for a wide variety of reasons including, but not limited to, research, mitigation for dam construction, supporting Tribal fishing rights, restocking of sport fishing stocks, and protection of endangered species. Commercial sales are not the primary reason these facilities operate. Examples of noncommercial facilities include Federal and State hatcheries, academic/research, Alaska nonprofit, and Tribal operations. Noncommercial facilities do not materially operate in a market economy nor are they required to generate balance sheets according to generally accepted principles. EPA, therefore, limited its information request in its detailed questionnaire to income sources, operating budgets, and production.

2.3.1   Facility Counts

Government facilities (Federal and State hatcheries) form the majority of in-scope noncommercial facilities. EPA estimates that there are approximately 141 noncommercial facilities with production greater than 100,000 lbs/yr. As explained in Section A.1, detailed questionnaire data are available for an estimated 139 noncommercial facilities. If the individual observations in the data set are multiplied by summary weights and all of the estimated counts are assigned to the classification of the observation that completed the detailed questionnaire, there are approximately 33 Federal facilities and 106 State facilities. Tribal facilities and the estimated impacts on them are assumed to be similar to these other noncommercial facilities. Alaska nonprofit facilities are not included in the presentation of detailed questionnaire data.

2.3.2   Production

Table 2-11 summarizes EPA's estimates of 2001 production from Federal and State facilities. Federal hatcheries are the largest in terms of average production at more than 430,000 lbs/yr, while State hatcheries, on average, produce 280,000 lbs/yr. All together, noncommercial facilities within the scope of the rule produced about 43.3 million pounds in 2001 (Table 2-11). Comparing information in Table 2-11 with that in Table 2-6, EPA estimates that there are 1.37 noncommercial facilities per commercial facility (i.e., 101 commercial facilities and 141 noncommercial facilities), but each noncommercial facility produces about 44 percent of a typical commercial facility (310,000 pounds to 700,000 pounds). There are several reasons for the differences between commercial and noncommercial facilities. Noncommercial facilities focus on getting a single crop in the water at a specific point in time while commercial facilities stagger their production to make bring in income throughout the year. Some noncommercial facilities focus on restoring endangered species and limit production to the capacity of the water body. Commercial facilities seek to maximize production for a given set of fixed costs. Commercial production figures also include the large net pen systems not seen in the noncommercial sector.

Table 2-11
Estimated Production and Employment for In-Scope Noncommercial Facilities by Type

Production System	Number In-Scope Facilities	Estimated Employment (FTE)	Estimated 2001 Production (Million lbs)	2001 Average Facility Employment	2001 Average Production (Million lbs)
Federal	33	275	14.1	8	0.43
State	106	1,288	29.2	12	0.28
Total	139	1,563	43.3	11	0.31

Source: EPA estimates from detailed survey (USEPA, 2002b).

2.3.3   Employment

Table 2-11 shows employment information. Overall, the in-scope noncommercial sector accounts for an estimated 1,563 jobs, of which 1,288 belong to State facilities. A typical State facility has 12 employees, about 1.5 times that for a Federal hatchery. The average production for a State facility is 64 percent that of a Federal hatchery.

2.3.4   Funding Sources

Noncommercial facilities derive their funding from a variety of sources. Unlike commercial facilities, EPA found few noncommercial facilities that sell more than a small amount of their production. Instead, their production is released into lakes, streams, and rivers to replenish wild stocks of endangered species and game fish. This lack of sales among these facilities means that these organizations rely on other sources to fund their operations. A number of State and Federal facilities, especially ones located in the West, receive funding as mitigation projects for dams that obstruct the natural spawning of species such as salmon. Other sources of revenue reported by State facilities includes Federal grants, State general funds, and fishing licenses.

EPA's detailed survey of noncommercial facilities collected information on operating budgets and also requested that the respondent identify facility funding from fishing licenses, commercial fishing permits, vanity tags for vehicles, and special-purpose stamps. For the purpose of this analysis, EPA combined these funds under the general term "User Fees." User fees offer States a way to meet the incremental costs of added pollution control. This option is not available to Federal or Tribal facilities. Among the 106 state facilities, 58 (55 percent) reported some type of income from user fees during 2001.

Of those facilities reporting user fees, the fees on average funded 81 percent of the budget for the facilities.

2.4   REFERENCES

ERG (Eastern Research Group). 2004. "Updated: Concentrated Aquatic Animal Production Industry: Unpaid Labor." Memorandum to Chris Miller, EPA, from ERG. February 9.

SBA (Small Business Administration). 2001. 13 CFR Parts 107 and 121 Size eligibility requirements for SBA financial assistance and size standards for agriculture. Direct Final Rule. 65 FR 100:30646-30649.

NMFS (National Marine Fisheries Service). 2003. *Fisheries of the United States: 2002.* U.S. Department of Commerce. National Oceanic and Atmospheric Administration. Silver Spring:MD. September.

USDA (U.S. Department of Agriculture, National Agricultural Statistics Service). 2000. *1998 Census of Aquaculture.* Also cited as 1997 Census of Agriculture. Volume 3, Special Studies, Part 3. AC97-SP-3. February.

USEPA (U.S. Environmental Protection Agency). 2004. Development Document for Final Effluent Limitations Guidelines and Standards for the Concentrated Aquatic Animal Production Industry Point Source Category. Washington, DC.

USEPA (U.S. Environmental Protection Agency). 2003. 40 CFR Part 451. Effluent Limitations Guidelines and New Source Performance Standards for the Concentrated Aquatic Animal Production Point Source Category; Notice of Data Availability; Proposed Rule. *Federal Register* 68:75068-75105. December 29.

USEPA (U.S. Environmental Protection Agency). 2002a. Economic and Environmental Impact Analysis of the Proposed Effluent Limitations Guidelines and Standards for the Concentrated Aquatic Animal Production Industry. Washington, DC. EPA-821-R-002-015. September.

USEPA (U.S. Environmental Protection Agency). 2002b. Detailed Questionnaire for the Aquatic Animal Production Industry. Washington, DC: OMB Control No. 2040-0240. Expiration Date November 30, 2004.

USEPA (U.S. Environmental Protection Agency). 2002c. Development Document for Proposed Effluent Limitations Guidelines and Standards for the Concentrated Aquatic Animal Production Industry Point Source Category. Washington, DC. EPA-821-R-02-016. August.

USEPA (U.S. Environmental Protection Agency). 2001. Screener Questionnaire for the Aquatic Animal Production Industry. OMB Control Number 2040-0237. Washington, DC. July.

USEPA (U.S. Environmental Protection Agency). 2000. *Guidelines for Preparing Economic Analyses.* Washington, DC: U.S. Environmental Protection Agency. EPA 240-R-00-003. September. DCN 20435.

CHAPTER 3

ECONOMIC IMPACT METHODOLOGY

This section provides an overview of the methodology used in the economic impact analysis to evaluate the effects of EPA's final regulation on commercial and noncommercial aquaculture facilities.

Section 3.1 presents EPA's cost annualization model that calculates the present value and annualized costs that feed into the other analyses. Sections 3.2 and 3.3 address EPA's approach to evaluate impacts to existing commercial and noncommercial facilities. Section 3.4 presents EPA's decision matrix for evaluating economic achievability for the final regulation.

Figure 3-1 illustrates the relationship among the different components of the analysis of commercial facilities. EPA's closure analysis compares the post-tax present value of earnings with and without incremental pollution control costs. Other direct impacts on employment and output are calculated from projected closures as a result of the rule. Additional commercial impacts are assessed using pre-tax annualized costs in a sales test and credit test, and using unannualized capital costs and output from the closure model to evaluate the financial health of the facility. EPA also follows the projected direct impacts from the closure analysis as they expand to affect local communities and the nation.

Figure 3-2 illustrates the relationship among the components for the economic impact analysis of noncommercial facilities. Since there are no tax considerations for these facilities, there is a simplified list of inputs to EPA's cost annualization model. The pre-tax annualized costs are used in a budget test and, where applicable, a user fee test.

Section 3.5 discusses EPA's approach to estimate whether the final rule presents a barrier to entry for new businesses. Finally, Section 3.6 summarizes the market characteristics of the U.S. aquaculture industry.

3.1 COST ANNUALIZATION MODEL

The starting point for the analyses presented in this report is EPA's cost annualization model. The model calculates four types of compliance costs: (1) present value of expenditures, before-tax basis; (2) present value of expenditures, after-tax basis; (3) annualized cost, before-tax basis; and (4) annualized cost, after-tax basis. The cost annualization model for this final regulation follows the methodology described for other effluent guidelines (e.g., see Concentrated Animal Feeding Operations (USEPA, 2002a, Appendix A)). This section provides an overview of the cost annualization model EPA uses for this analysis.

3.1.1 Input Data Sources

The cost annualization model requires several key data inputs for estimating annual costs of compliance with the rule:

Figure 3-1
Commercial Facilities: Economic Analysis Flowchart

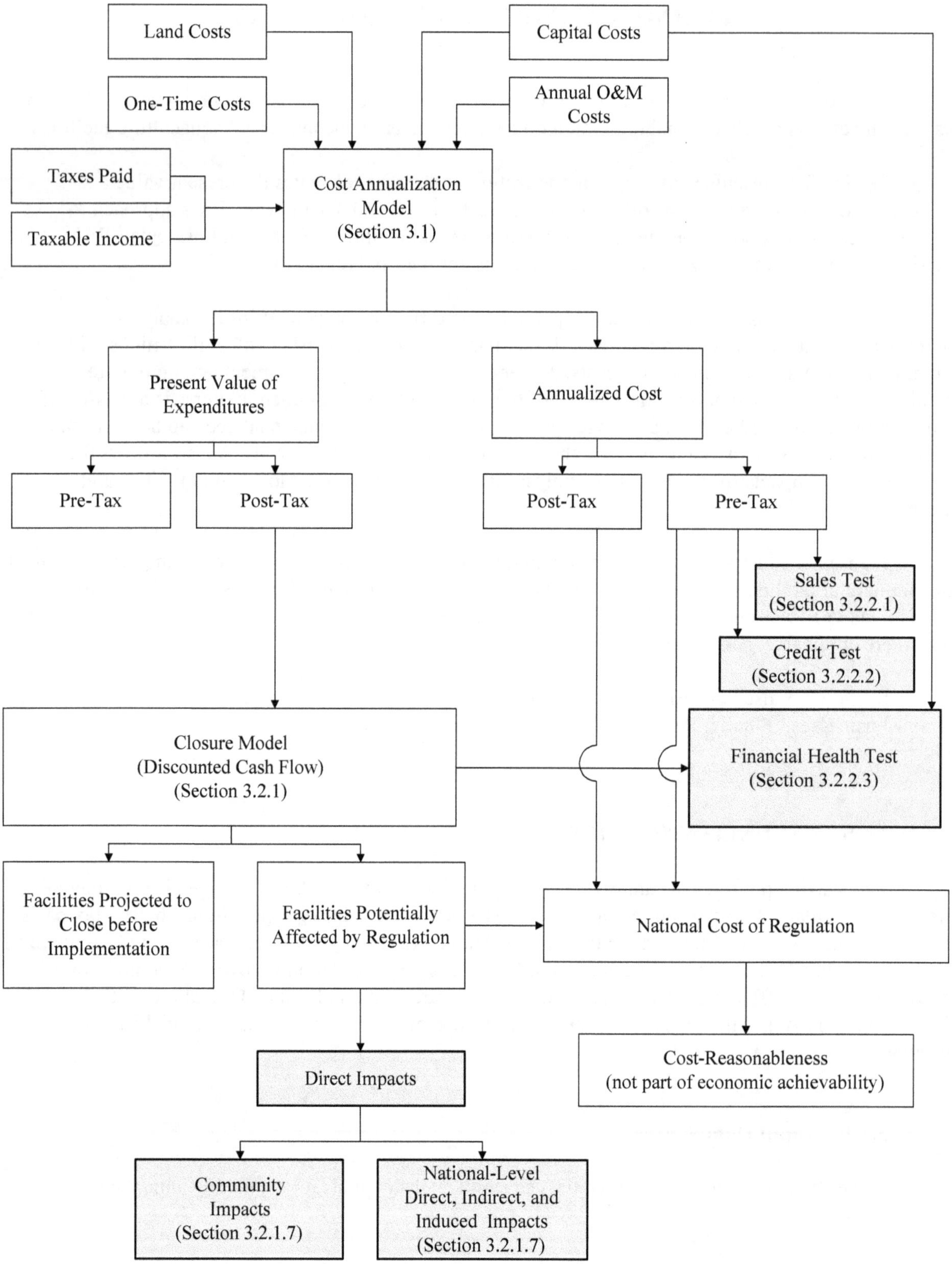

Figure 3-2
Noncommercial Facilities: Economic Analysis Flowchart

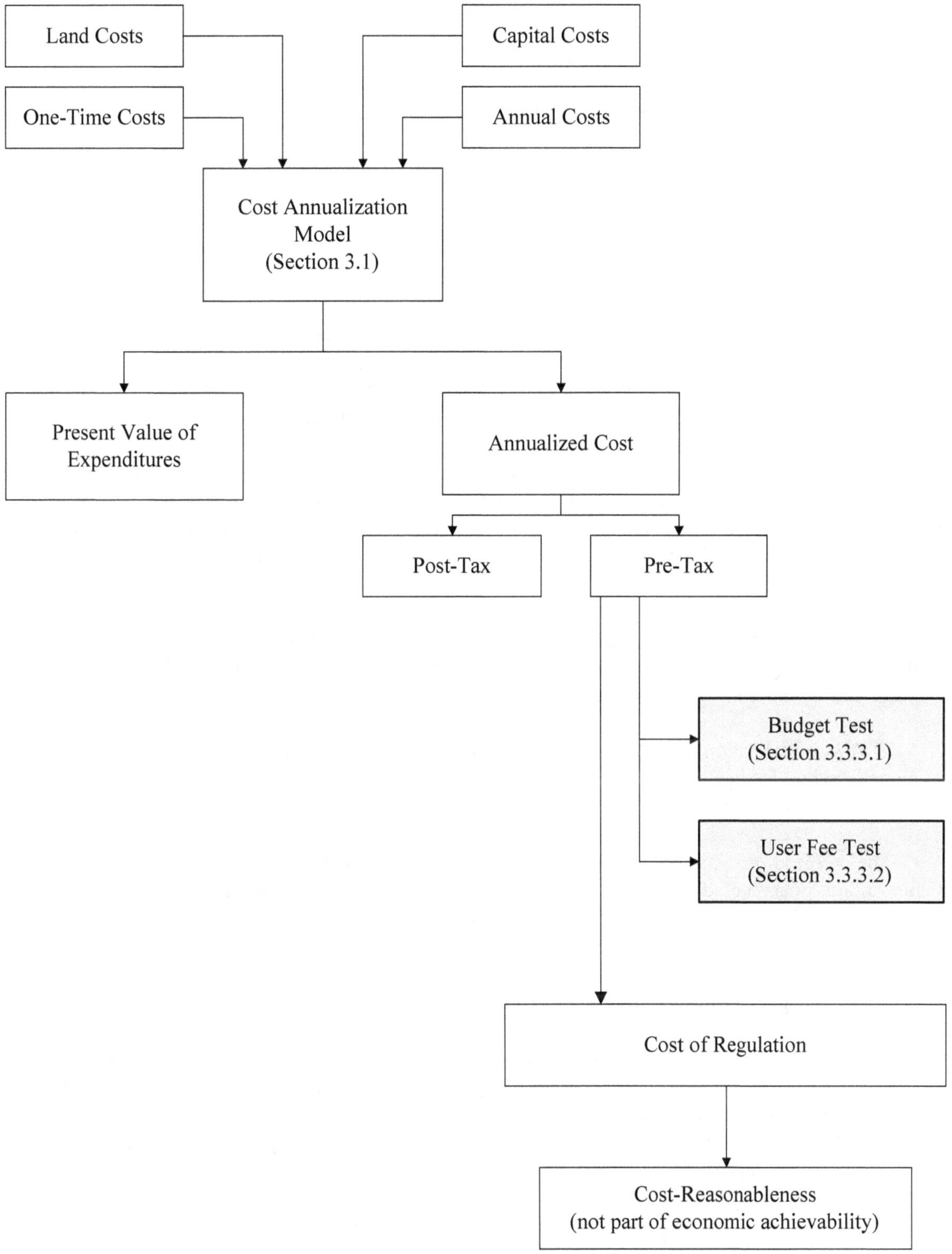

- ▪️ Land costs

- ▪️ Capital costs

- ▪️ One-time non-capital costs

- ▪️ Annual operating and costs (O&M)

- ▪️ Depreciable life of the asset

- ▪️ Discount rate

- ▪️ Marginal tax rate (Federal and State)

EPA's *Development Document* for the final rule provides detailed information on how the Agency developed the *land, capital, one-time non-capital, and O&M* costs that are input in the cost annualization model (USEPA, 2004a). The land cost reflects either the actual cost of purchasing land for additional pollution control measures or the opportunity value of land already owned but must now be put aside for incremental pollution control measures. Land costs are one-time costs that cannot be depreciated. The capital cost is the initial investment needed to purchase and install the structure; it is a one-time cost which can be depreciated. One-time non-capital costs, such as an engineering report cannot be depreciated. The O&M cost is the annual cost of operating and maintaining the incremental pollution control measures. The maintenance component includes capital replacement costs to keep the system running.

The *depreciable life of the asset* refers to EPA's assumption of the time period used to depreciate capital improvements that are made because of the rule. The cost annualization model uses a 10-year period and a mid-year convention (see Section 3.1.2).

EPA's annualization model uses a real *discount rate* of 7 percent, as recommended by the Office of Management and Budget (OMB, 2003). EPA assumes this input to be a real interest rate, and therefore it is not adjusted for inflation.

The *marginal tax rate* used to compute the tax shield depends on the amount of taxable earnings (revenues minus costs including interest) at the regulated entity. Inputs to the cost annualization model to calculate a facility's tax shield include both Federal and State tax rates.

3.1.2 Depreciation Method

EPA examined three alternatives to depreciate capital investments, including Modified Accelerated Cost Recovery System (MACRS), straight-line depreciation, and section 179 of the Internal Revenue Code (USEPA, 2002a). EPA chose to use the MACRS which allows businesses to depreciate a higher percentage of an investment in the early years and a lower percentage in the later years. In contrast, straight-line depreciation writes off a constant percentage of the investment each year. MACRS offers companies a financial advantage over the straight-line method because an aquaculture facility's taxable income may be reduced under MACRS by a greater amount in the early years when the time value of money is greater. EPA also considered using the Internal Revenue Code Section 179 provision to elect to expense up to $100,000 in the year the investment is placed in service, assuming that the investment costs do not exceed $400,000 (PL 108-27, 2003 and IRS, 2003). EPA assumes, however, that this provision is already applied to other investments at the facility.

Under MACRS, the cost of property is recovered over a set period. The recovery period is based on the property class to which your property is assigned. To determine the recovery period of depreciable property, IRS identifies asset classes based on the activity for which the property is being used. EPA has identified the appropriate class for each type of cost and has judged that a 10-year time frame is appropriate for the economic analysis supporting the CAFO regulation and this final regulation. More information is provided in Appendix A of the CAFO rule, (USEPA, 2002a). A 10-year depreciation time frame is consistent with the 10-year property classification of a single-purpose livestock structure, which is defined under IRS Section 168(i)(13)(B) as any enclosure or structure specifically designed, constructed and used for housing, raising, and feeding a particular kind of livestock, including their produce, or for housing the equipment necessary for the housing, rasing, and feeding of livestock (IRS, 1999).

EPA uses a mid-year convention for calculating depreciation. This means that EPA assumes that the capital investment is made at the beginning of the year and the facility goes into operation six months later.

3.1.3 Tax Rates

The cost annualization model uses both Federal and State tax rates as inputs to calculate an average commercial operation's tax shield. (Noncommercial operations have no tax shield). EPA calculated national average state income tax rates of 6.6 percent for corporations and 5.8 percent for individuals (Table 3-1, taken from USEPA, 2002a) and these rates are used in this analysis. Depending upon the survey response, the model calculates the tax shield based on the corporate tax rate for C corporations, personal tax rate for partnerships and proprietorships, and no tax rate for S corporations or Limited Liability Corporations. EPA uses the net present value of after-tax costs for the closure analysis because it reflects the impact the business would actually see in its earnings.

Table 3-1
State Income Tax Rates

State	Corporate Income Tax Rate	Basis for States With Graduated Tax Tables	Personal Income Tax Upper Rate	Basis for States With Graduated Tax Tables
Alabama	5.00%		5.00%	$3,000+
Alaska	9.40%	$90,000+	0.00%	
Arizona	9.00%		6.90%	$150,000+
Arkansas	6.50%	$100,000+	7.00%	$25,000+
California	9.30%		11.00%	$215,000+
Colorado	5.00%		5.00%	
Connecticut	11.50%		4.50%	
Delaware	8.70%		7.70%	$40,000+

State	Corporate Income Tax Rate	Basis for States With Graduated Tax Tables	Personal Income Tax Upper Rate	Basis for States With Graduated Tax Tables
Florida	5.50%		0.00%	
Georgia	6.00%		6.00%	$7,000+
Hawaii	6.40%	$100,000+	10.00%	$21,000+
Idaho	8.00%		8.20%	$20,000+
Illinois	4.80%		3.00%	
Indiana	3.40%		3.40%	
Iowa	12.00%	$250,000+	9.98%	$47,000+
Kansas	4.00%	$50,000+	7.75%	$30,000+
Kentucky	8.25%	$250,000+	6.00%	$8,000+
Louisiana	8.00%	$200,000+	6.00%	$50,000+
Maine	8.93%	$250,000+	8.50%	$33,000+
Maryland	7.00%		6.00%	$100,000+
Massachusetts	9.50%		5.95%	
Michigan	2.30%		4.40%	
Minnesota	9.80%		8.50%	$50,000+
Mississippi	5.00%	$10,000+	5.00%	$10,000+
Missouri	6.25%		6.00%	$9,000+
Montana	6.75%		11.00%	$63,000+
Nebraska	7.81%	$50,000+	6.99%	$27,000+
Nevada	0.00%		0.00%	
New Hampshire	7.00%		0.00%	
New Jersey	7.25%		6.65%	$75,000+
New Mexico	7.60%	$1 Million+	8.50%	$42,000+
New York	9.00%		7.88%	$13,000+
North Carolina	7.75%		7.75%	$60,000+
North Dakota	10.50%	$50,000+	12.00%	$50,000+
Ohio	8.90%	Based on Stock Value	7.50%	$200,000+
Oklahoma	6.00%		7.00%	$10,000+
Oregon	6.60%		9.00%	$5,000+
Pennsylvania	9.90%	1997 and thereafter	2.80%	
Rhode Island	9.00%		10.40%	$250,000+
South Carolina	5.00%		7.00%	$11,000+
South Dakota	0.00%		0.00%	
Tennessee	6.00%		0.00%	
Texas	0.00%		0.00%	
Utah	5.00%		7.20%	$4,000+
Vermont	8.25%	$250,000+	9.45%	$250,000+

State	Corporate Income Tax Rate	Basis for States With Graduated Tax Tables	Personal Income Tax Upper Rate	Basis for States With Graduated Tax Tables
Virginia	6.00%		5.75%	$17,000+
Washington	0.00%		0.00%	
West Virginia	9.00%		6.50%	$60,000+
Wisconsin	7.90%		6.93%	$20,000+
Wyoming	0.00%		0.00%	
Average:	6.61%		5.84%	

Source: CCH, 1999a and 1995.

Basis for rates is reported to nearest $1,000. Personal income tax rates for Rhode Island and Vermont based on federal tax (not taxable income). Tax rates given here are equivalents for highest personal federal tax rate.

3.1.4   Tax Shield Not Included

The cost annualization model does not consider tax shields on interest paid to finance incremental pollution control. The cost annualization model assumes a cost to the operation to use the money (the discount/interest rate), whether the money is paid as interest or is the opportunity cost of internal funding. Tax shields on interest payments are not included in the cost annualization model because it is not known what mix of debt and capital an operation will be used to finance the cost of incremental pollution control and to maintain a conservative estimate of the after-tax annualized cost.

3.1.5   Sample Cost Annualization Spreadsheet

Table 3-2 shows a sample cost annualization worksheet. The same worksheet is used for commercial and noncommercial facilities but with different tax effects. The top of the spreadsheet shows the data inputs described in Section 3.1.1. For the example, sample data for a fictitious survey identification number "XYZ" is read into the calculations. The assumed land, capital, annual O&M, and one-time costs are $1,000, $2,000, $100, and $10, respectively. The facility belongs to a corporation that had earnings before taxes (EBT) of $15,000 in 2001 and paid an average of $700 in annual taxes over 1999-2001. (EBT and average taxes are calculated from detailed questionnaire data for the actual analysis.) The model uses a mid-year convention and this effect will be seen most clearly in Year 1 and Year 11.

Table 3-2
Concentrated Aquatic Animal Production Cost Annualization Model

INPUTS

Survey ID #:
Option Number:

	2001	Engineering Inputs 2001 1	Economic Analysis 2001 1
	XYZ		

	2001	Year Dollars	ENR CCI
Land Cost	1000	$1,000.0	
Initial Capital Cost	$2,000.0	$2,000.0	
Annual Operation & Maintenance Cost	$100.0	$100.00	
One-Time Non-Equipment Cost	$10.0	$10.0	
Real Discount Rate	7.0%		
Corporate Tax Structure	1		
EBT :	$15,000.0		
Taxes Paid (3-yr average):	$700.0		
Marginal Income Tax Rates:			
Federal	15.0%		
State	6.6%		
Combined	21.6%		

Federal Corp. Tax Table:

Taxable Income ($)	Marginal Tax Rate
$0	15.0%
$50,000	25.0%
$75,000	34.0%
$100,000	34.0%
$10,000,000	35.0%

Federal Personal Tax Table:

Taxable Income ($)	Marginal Tax Rate
$0	15.0%
$25,700	28.0%
$62,450	31.0%
$130,250	36.0%
$250,000	39.6%

Column 1	2	3	4	5	6	7	8	9
Year	Depreciation Rate	Depreciation For Year	Tax Shield From Depreciation	O&M Cost	O&M Tax Shield	Cash Outflow	Adjusted Tax Shield	Cash Outflow After Tax Shields
1	10.00%	$200	$43	$60	$13	$3,060	$56	$3,004
2	18.00%	$360	$78	$100	$22	$100	$99	$1
3	14.40%	$288	$62	$100	$22	$100	$84	$16
4	11.52%	$230	$50	$100	$22	$100	$71	$29
5	9.22%	$184	$40	$100	$22	$100	$61	$39
6	7.37%	$147	$32	$100	$22	$100	$53	$47
7	6.55%	$131	$28	$100	$22	$100	$50	$50
8	6.55%	$131	$28	$100	$22	$100	$50	$50
9	6.56%	$131	$28	$100	$22	$100	$50	$50
10	6.55%	$131	$28	$100	$22	$100	$50	$50
11	3.28%	$66	$14	$50	$11	$50	$25	$25
Sum	100.00%	$2,000	$432	$1,010	$218	$4,010		$3,360
Present Value		$1,572	$339	$737	$159	$3,737		$3,238

	Before Tax Shield	After Tax Shield
Present Value of Incremental Costs:	$3,737	$3,238
Annualized Cost:	$498	$432

Notes: This spreadsheet assumes that a modified accelerated cost recovery system (MACRS) is used to depreciate capital expenditures.
Depreciation rates are from 1995 U.S. Master Tax Guide for 10-year property and mid-year convention.
First Year is not discounted.

3-8

The spreadsheet contains numbered columns that calculate the before-and after-tax annualized cost of the investment. Column 1 of Table 1 lists each year of the investment's life span, from its installation through its 10-year depreciable lifetime (shown over years 1 through 11 because a mid-year convention is used).

Column 2 represents the percentage of the capital costs that can be written off or depreciated each year. These rates are based on the MACRS and are taken from the *2000 U.S. Master Tax Guide* (CCH, 1999b). Multiplying these depreciation rates by the capital cost gives the annual amount the operator may depreciate, which is listed in Column 3. In the example, the capital expense results in $200 in depreciation in Year 1 ($2001). EPA uses depreciation expense to offset annual income for tax purposes; Column 4 shows the tax shield provided from the depreciation expense—the overall tax rate times the depreciation amount for the year. The corporation is in the 15 percent Federal tax bracket and a 6.6 percent State tax bracket. The depreciation tax shield is $43 dollars ($200 multiplied by 21.6 percent).

Column 5 is the annual O&M expense plus the one-time non-capital costs. Because of the mid-year convention assumption for depreciation, Year 1 and Year 11 show only 6 months of annual O&M costs or $50. Year 1 O&M also includes the one-time costs of $10 for a total of $60. Years 2 through 10 include annual O&M. Column 6 is the tax shield or benefit provided from expensing the O&M costs.

Column 7 lists the annual cash outflow, or total expenses, associated with the incremental pollution control under the analysis assumptions presented here. Total expenses include land, capital, one-time, and six months of O&M costs (i.e.,$1,000 plus $2,000 plus $10 plus $50 or $3,060).

Column 8 adjusts the sum of entries in Columns 4 and 6 to limit the projected tax shields to the average amount of taxes paid each year. In this example, the corporation has been paying an average of $700 in taxes, so the tax shield shown in Column 8 is not limited. Column 9 is the cash outflow after tax shields (i.e., Column 7 minus Column 8).

In the lower part of Table 3-2, the sum of the depreciation percentages is 100 percent, the sum of the depreciation taken is $2,000, total O&M and one-time costs are $1,010. Total cash outflow over the 10-year period is $4,010, which drops to $3,360 after the tax shields are considered.

EPA calculates the present value (NPV) of the cash outflows as:

where:
$$NPV = v_1 + \sum_{i=2}^{N} \frac{v_i}{(1+r)^{i-1}}$$

$v_1...v_n$ = series of cash flows
r = interest rate
n = number of cash flow periods
i = current iteration

EPA transforms the present value of the cash outflow into a constant annual payment for use as the annualized compliance cost. Columns 7 and 9 calculate the annualized cost as a 10-year annuity that has the same present value as the total cash outflow. The annualized cost represents the annual payment required to finance the cash outflow after tax shields. In essence, paying the annualized cost each year and paying the amounts listed in Columns 7 or 9 for each year are equivalent. EPA calculates the annualized cost as follows (where n is the number of payment periods):

$$\text{Annualized Cost} = \text{present value of cash outflows} * \frac{\text{real discount rate}}{1 - (\text{real discount rate} + 1)^{-n}}$$

In the Table 3-2 example, the annualized cost is $432 after accounting for the estimated tax shields and $498 without tax shields. There are two ways to calculate post-tax annualized cost. One way is to calculate the annualized cost as the difference between the annuity value of the cash flows (Column 7) and the tax shields (Columns 4 and 6). The second way is to calculate the annuity value of the cash flows after tax shields (Column 9). Both methods yield the same result.

EPA uses the pre-tax annualized cost to calculate the total social cost of the regulation (see Table 4-3) used in the cost-effectiveness and cost-reasonableness calculations. This approach incorporates the cost to industry for the purchase, installation, and operation of additional pollution control as well as the cost to Federal and State government from lost tax revenues. (Every tax dollar that a business does not pay due to a tax shield is a tax dollar lost to the government.) Note also that operation and maintenance costs include the cost of capital replacement. That is, if a component has a 5-year lifetime, the cost estimates for the 10-year period include the costs for two components.

EPA uses the post-tax annualized cost to reflect what a business actually pays to comply with incremental pollution control requirements. The post-tax present value of incremental pollution control costs is subtracted from the present value of forecasted earnings (2005-2015) to calculate post-regulatory value of future earnings in the closure analysis at the enterprise, facility, and company levels. See Section 3.2 for a discussion of the closure analysis and earnings forecast. For noncommercial operations, EPA's analysis assumes there is no difference between the pre-tax and post-tax estimates (i.e., noncommercial operation do not incur Federal or State tax costs).

3.2   COMMERCIAL FACILITIES

3.2.1   Closure Analysis

EPA developed a financial model to estimate whether the additional costs of complying with the final regulation rendered a regulated operation unprofitable. If so, the operation is projected to close as a result of the regulation, leading to impacts such as losses in employment and revenue. This financial model is also referred to as the "closure model" within this report. The closure analysis is performed at three levels:

- Enterprise (where aquaculture is only one of multiple operations at the farm/facility);

- Facility;

- Company (which may operate more than one facility).

For the sake of simplicity, the rest of the discussion of the closure model will use only the term "facility." The model is based on data from the detailed questionnaire (USEPA, 2002b). Facility-specific pairs of cost and revenues for actual aquaculture operations are not available elsewhere.

The closure model uses data and methodologies available to corporate financial analysts. The model compares future earnings with and without the regulation. The closure decision is modeled as:

Post-regulatory status = Present value of future earnings minus
the present value of after-tax incremental pollution control costs

The model projects the long-term effects of added pollution control costs on earnings. If the post-regulatory status is zero or less, the facility is projected to close.[1] Although simple in concept, the model incorporates numerous assumptions, including:

- ◻ How to calculate earnings (e.g., discounted cash flow or net income)

- ◻ Forecasting method(s) for future earnings as determined by prices;

- ◻ Time frame for consideration;

- ◻ The ability of the industry to pass costs through to consumers; and

- ◻ Discount rate (cost of capital)

The question of how to calculate earnings entails a series of other topics, such as cash flow, net income, depreciation, sunk costs, capital replacement and unpaid labor and management. Appendix A contains the detailed discussion required for each of the topics pertaining to earnings.

The rule is scheduled to be promulgated in 2004, so the time frame for the projection is from 2005-2015. EPA's closure analysis therefore compares earnings during 2005-2015 with and without cost of compliance under the final regulation. EPA uses two methods to estimate earnings for the purposes of its closure analysis: cash flow and net income. The difference between the cash flow and net income calculations is depreciation (a non-cash cost). Depreciation is included as a cost in the net income basis but not in the cash flow basis.

For the purposes of this analysis, EPA calculates the difference between gross revenues and total expenses reported in the detailed questionnaire and reduced the value by the estimated Federal and State taxes to calculate net income. EPA then adds the non-cash expense of depreciation (when it was reported in the questionnaire) to net income to calculate cash flow. This approach is consistent with the guidance from the Farm Financial Standards Council (FFSC, 1997) and several business financial references (Brealy and Meyers, 1996; Brigham and Gapenski, 1997; and Jarnagin, 1996). As part of this analysis, EPA examines the possibility of closure under three forecasting methods to project future earnings (see Section 3.2.1.1 below). EPA's forecasting model, like the cost annualization model, uses a real discount rate of 7 percent, as recommended by the Office of Management and Budget (OMB, 2003). EPA assumes this input to be a real interest rate, and therefore it is not adjusted for inflation.

For the purposes of assessing economic achievability, EPA assumes that the costs of the rule are not passed on to consumers. The facility must absorb all increased costs. If it cannot do so and remain in

---

[1] EPA assumes that it no longer operates and that closure-related impacts result. In contrast, facilities that are sold because a new owner presumably can generate a greater return are considered *transfers*. Transfers cause no closure-related impacts, even if the transfer was prompted by increased regulatory costs. Transfers are not estimated in this analysis.

operation, all production is assumed lost. EPA's assumption of no cost pass through is a more conservative approach to evaluating economic achievability among regulated entities. In addition, EPA's closure analysis does not incorporate the ability of other facilities to increase production to offset the closure. More information on EPA's rationale for this approach is provided in Section 3.6. (To evaluate market and trade level impacts, however, EPA assumes all costs are shifted onto the broader market level as a way of assessing the upper bound of potential effect; see Section 5.3.)

3.2.1.1 Data Sources for Forecasting Methods

EPA examined four sets of data from various Federal agencies as possible bases for forecasting future earnings for concentrated aquatic animal production facilities. The first set is the agricultural baseline projections developed by USDA. The other three sets of data are historical price data collected by the U.S. Department of Labor, Bureau of Labor Statistics, USDA, and the National Marine Fisheries Service. Each data set is discussed in a separate section below.

__USDA Agricultural Baseline Projections__. USDA provides long-run baseline projections for the agricultural sector annually through 2013 (USDA, 2003a and USDA, 2004a). In the chapter on U.S. agricultural sector aggregate indicators, the report presents projections for farm income, food prices, and U.S. trade value. The USDA model projects a strengthening in economic growth which results in rising market prices and farm income as well as an improvement in the financial condition of the agricultural sector (USDA, 2003a, page 65; USDA, 2004a, page 62). Table 3-3 and Figure 3-3 reproduce the data from USDA, 2003a, Table 31 and USDA, 2004a, Table 31 for the Consumer Price Index, Food at Home, Fish and Seafood sector for 2000 through 2013 (which includes canned tuna and frozen fish). The data for 2000 through 2002 indicate a downturn consistent with the data from the sources described below and the downturn mentioned in comments by the Joint Subcommittee on Aquaculture (JSA; JSA, 2003). Forecasts begin with a 0.9 percent recovery in 2003 and a 2.5 percent annual increase from 2004 through 2013. Because USDA interprets the rising consumer price index to translate into improved farm income and financial condition, EPA assumes that the prices are rising at a greater rate than costs and that this index might provide a basis for forecasting future earnings for aquatic animal production facilities. EPA also assumes that the market pressures for wild and farmed fish and seafood are comparable due to the substitutability of the products. The USDA projections reflect the 2000-2002 downturn but assume that conditions return to a long-term upward trend such as that visible in the previous decades. However, EPA does not know if a long-term upward trend will return to fish prices. Therefore EPA's forecasting approach incorporates forecasts, using additional sources of data, that do not show long-term upward trends.

Table 3-3
USDA Consumer Food Price Index-Food at Home, Fish and Seafood

	Year													
	2000	2001	2002	2003	2004	2005	2006	2007	2008	2009	2010	2011	2012	2013
Index	190.4	191.1	188.1	189.8	194.5	199.4	204.4	209.5	214.7	220.1	225.6	231.2	237.0	242.9
%Chg	2.8	0.4	-1.6	0.9	2.5	2.5	2.5	2.5	2.5	2.5	2.5	2.5	2.5	2.5

Note: 1982-1984 = 100
Source: USDA, 2003a and 2004a.

__U.S. Department of Labor, Bureau of Labor Statistics (BLS), Producer Price Index.__ The U.S. Department of Labor, Bureau of Labor Statistics' Producer Price Index (PPI) are monthly estimates of the average change over time in the selling prices received by domestic producers for their output. The prices included in the PPI are from the first commercial transaction for many products and some services. EPA downloaded two time series from the BLS website for analysis:

  ■▯ Unprocessed and packaged fish ("Fish PPI")
      Not seasonally adjusted
      Series ID: WPU0223
      1980:1-2003:12

  ■▯ Shrimp ("Shrimp PPI")
      Not seasonally adjusted
      Series ID: WPU02230501
      1991:1-2003:12

EPA also downloaded Series ID WPU02230101, Salmon, not seasonally adjusted, 1980:1-2002:9. No trend was identified in the data.

  Figure 3-4 illustrates the monthly PPI for unprocessed and packaged fish for the period of 1980 through 2004 (BLS, 2004a). Visual inspection, by itself, indicates how prices in the last few years in the sector show a different pattern of behavior than in the preceding decades. This hypothesis is tested below. Figure 3-5 shows monthly PPI for shrimp from 1991 through 2003 (BLS, 2004b). Although the time period is shorter than that seen in Figure 3-4, the recent downturn in producer prices is just as evident.

  __USDA Trout Price Data__. EPA downloaded annual average prices for trout 12 inches or longer from 1994 through 2002 from the USDA web site. These data are reported in USDA's series *Trout Production* (USDA, 2003b). These data are converted to constant dollars using the Consumer Price Index CPI-U (CEA, 2004, Table B-60), see Figure 3-6.[1] The downward trend in producer prices is evident.

  __National Marine Fisheries Service Price Data__. EPA examined a second set of trout price data from the National Marine Fisheries Service, which collects price data for various species and markets. EPA downloaded the monthly average price for fresh boned Idaho trout at the Fulton Fish Market from 1990:1 through 2002:12 (NMFS, 2004). As with the USDA trout data, prices are converted to constant dollars using the Consumer Price Index CPI-U (CEA, 2004). Figure 3-7 shows a pronounced downward trend in prices in recent years leveling off after March 2001.

  __USDA Trade Adjustment for Farmers, Preferred Price Data.__ EPA examined the data sources used to support a relatively new USDA program. In August 2002, Congress enacted the Trade Act of 2002 (Public Law 107-210). The Act establishes a new program—Trade Adjustment Assistance for Farmers (TAA). The program is administered by USDA's Foreign Agricultural Service (FAS). FAS published a final rule implementing the program in August 2003 (USDA, 2003c).

---

  [1]To covert data to a specific year, multiply the datum by the ratio of the CPI-U values for the appropriate years. For example, CPI-U (2001) = 179.1 and CPU-U (1994) = 148.2; $100 in 1994 dollars converts to $120.9 in 2001 dollars (i.e., 100 * 179.1/ 148.2).

Under the program, a group of at least three agricultural commodity producers submits a petition to FAS. FAS reviews the petition and, if acceptable, publishes a notice in the *Federal Register* that the petition has been received. Another part of USDA, the Economic Research Service (ERS), conducts a market study to verify the decline in prices, the potential impact of imports, etc., and reports the findings back to FAS. FAS then determines whether the petitioners are eligible for trade adjustment assistance. Assistance takes two forms: technical expertise and cash benefits. More information of the program is available in the rulemaking record (ERG, 2004a).

FAS maintains a registry for petitions at http://www.fas.usda.gov/itp/taa/registry.htm. As of January 2004, FAS had received 11 petitions from salmon, shrimp, catfish and crayfish farmers in the U.S. Of these petitions, 7 had been approved as eligible for trade adjustment assistance. With the exception of some of the denied petitions, price data submitted with petitions are not publicly available. EPA follows FAS methodology regarding the order of preference for data sources for the purposes of developing its forecasting methods:

- The preferred source of data are from USDA National Agricultural Statistics Service (NASS). Therefore, EPA's forecasts use NASS price data for trout, a major commodity considered under the rule.

- EPA and FAS use National Marine Fisheries Service (NMFS) data when NASS data are not available. EPA examined both NASS and MMFS trout price data when developing its forecasts.

- TAA petitions examine six years of data (the current year and the previous five years). EPA projections are made on the basis of time series data that reflect 24, 12, and 9 years of data for Fish PPI, Shrimp PPI, and NASS Trout prices, respectively.

EPA's forecasting model, therefore, broadly adheres to USDA's approach under its Trade Adjustment Assistance for Farmers program. EPA uses USDA's preferred source for price data (i.e., NASS) and uses longer time series (9 to 12 years compared to TAA's use of six years of data).

3.2.1.2 Forecasting Methods

The USDA agricultural baseline projections provide one forecasting method. Given the other data in Section 3.2.1.1, however, EPA decided to develop alternate forecasts to serve as counterpoints to USDA's projections for fish and seafood prices. Section 3.2.1.2 describes the process used to fit trends to the historical data.[2] Forecasts developed from these data, then, are simple extrapolations of recent trends.

All data are converted to constant dollars, where necessary. For time series data with monthly observations, EPA converts the series to a 12-month centered moving average to smooth away any seasonal variation. EPA identified no cyclicality in the data beyond seasonality.

Figure 3-8 shows the monthly Producer Price Index for unprocessed and packaged fish ("Fish PPI"). The thin jagged line is the raw data also shown in Figure 3-4. The smoothed data are shown by a

---

[2] All regressions were done in E-Views and are significant at a 0.01 level or better.

thin continuous line. The data series appears to have points in time where the slope of the trend seems to change. EPA conducted the Chow Breakpoint Test multiple times in the regions where the underlying slopes seemed to change in order to logically break the time series (Kennedy, 1998). The Chow Breakpoint Test compares the sum of squares between a restricted and unrestricted model. In the breakpoint test, the unrestricted model allows the slope coefficient to be different before and after the suggested breakpoint period. EPA tested a series of candidate breakpoints and selected the one with the largest F-statistic to serve as the point where the slope of the trend changed. A dummy variable indicating the later period of the trend and a crossproduct of the dummy and the time variable were added as independent variables to the trend regression to allow the latter part of the trend line to have a different intercept and slope than the earlier part. In this way, a simple ordinary least squares regression could estimate a kinked trend line.[3]

EPA's examination of the Fish PPI appear to exhibit three different slopes during the 23 year period so a middle set of dummies and cross-products was added allowing the trend line to have three different slopes. The kinked dotted line shown in Figure 3-8 is the fitted trend. That is, the trend line PPI for unprocessed and packaged fish appears to have three different slopes during the January 1980 through December 2003 period. As shown, the downward slope for the period after December 1999 levels off in 2002 and 2003.

The Chow Breakpoint Test was performed on the BLS shrimp data (Shrimp PPI), USDA trout data, and the NMFS Fulton Fish Market price data. Figure 3-9 is the counterpart to Figure 3-5; i.e., is shows the raw shrimp PPI data, the 12-month smoothed average, and the fitted trend. The price breakpoint for this species is at the end of 1997. Figures 3-10 and 3-11 show the two data sets for trout prices. The USDA data show a continual downward trend while the Fulton fish market data level off through 2001 through 2003.

3.2.1.3 Index For Use in Projecting Future Earnings

The trends fitted in Section 3.2.1.2 provide forecasting equations to project future price levels. The methods for forecasting future earnings are implemented with facility-specific questionnaire data. The price level forecasts are converted into an index with 2001 as the base period because this is the most recent year for which data were collected in the detailed questionnaire. The earnings forecast is assumed to begin in 2005 and end in 2015, coinciding with EPA's schedule for promulgating this final regulations.

Figure 3-12 illustrates the forecasting indices. The base year for the index is 2001 and the thin unbroken line is the 100 grid line. The solid line with the stars is USDA's baseline projection. EPA believes these forecasts are optimistic. USDA's projections are the only forecast EPA found that show prices increasing over the long term. The other forecasts show downward trends with the Shrimp PPI showing the steepest downward slope (see Figure 3-12). While all the forecasts are simple trend lines, the forecasts developed using historical fish prices recognize that the trend may have changed in the recent past. The usual caveats about reliance on simple trend projection apply, however, similar caveats apply to

---

[3]Other techniques are available to accomplish similar ends, such as spline regression and the use of squared or cubed terms to estimate a Taylor series equation, however, EPA was not concerned with assessing the continuity of the prediction function and had no reason to examine more than single-point changes in slope, i.e. kinks, rather than curves.

the official USDA forecast which appears to dismiss recent changes in the market place. In effect, Figure 3-12 suggests that, if the market for fish has changed, the industry may face for financial difficulties regardless of any potential costs of additional pollution control resulting from the rulemaking effort.

3.2.1.4 Selected Projection Methods for Future Earnings

The broad difference between the USDA projections and the other forecasting methods depends on whether or not recent changes in the marketplace are temporary or permanent. While more detailed modeling of the markets for various species might be feasible, the results would likely be within the range bounded by the USDA and EPA projections.

Another possibility is that the data collected in the detailed questionnaire for 1999 through 2001 reflect conditions among the surveyed businesses that are projected to continue into the future. That is, the future looks like the recent past and a 3-year average provides a naive baseline for projecting future earnings. Again, the projections using this method would likely result in estimates that are generally– but not always– between those of USDA and EPA projections.

For the purposes of this analysis, EPA uses three forecasting methods for its facility closure analysis. One forecasting method uses USDA projections, starting with 2001 earnings. A second forecasting method uses an EPA projection, starting with 2001 earnings. This approach incorporates the PPI for shrimp, USDA's trout prices, and the Fish PPI for all other species.[4] For example, if a facility raises a 50:50 mix of trout and another species, the forecast is based on a weighted average of the indices. The third forecasting method uses average of 1999-2001 earnings. The base year for the index is 2001, which is the starting year for the first two projection methods. Table 3-4 shows a list of the indices that EPA uses for its closure model.

---

[4]USDA price data are available for catfish and trout. Most catfish, however, are raised in ponds, which is a production method outside the scope of the rulemaking.

Figure 3-3
Annual USDA Consumer Food Price Index-Food at Home
Fish and Seafood

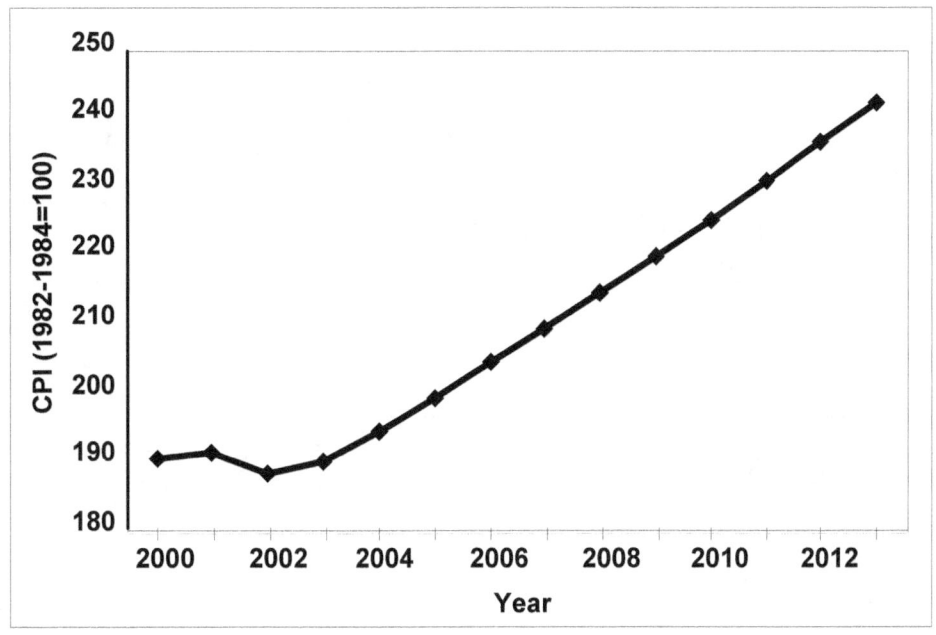

Source: USDA, 2003a and 2004a.

Figure 3-4
Unprocessed and Packaged Fish-Monthly PPI

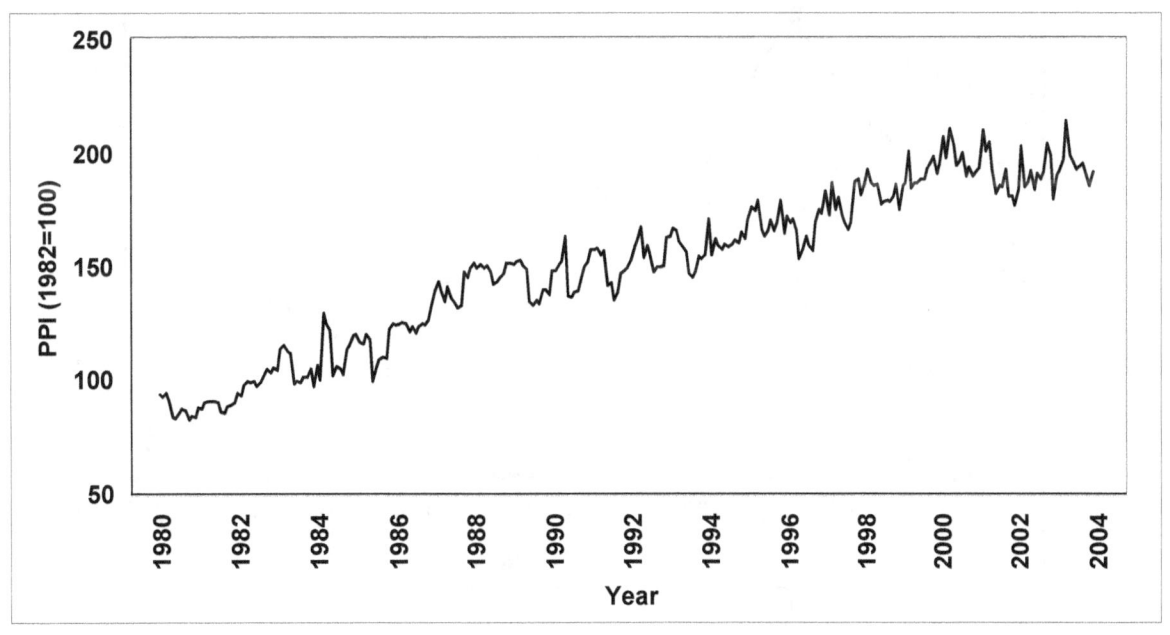

Source: BLS, 2004a.

Figure 3-5
Shrimp-Monthly PPI

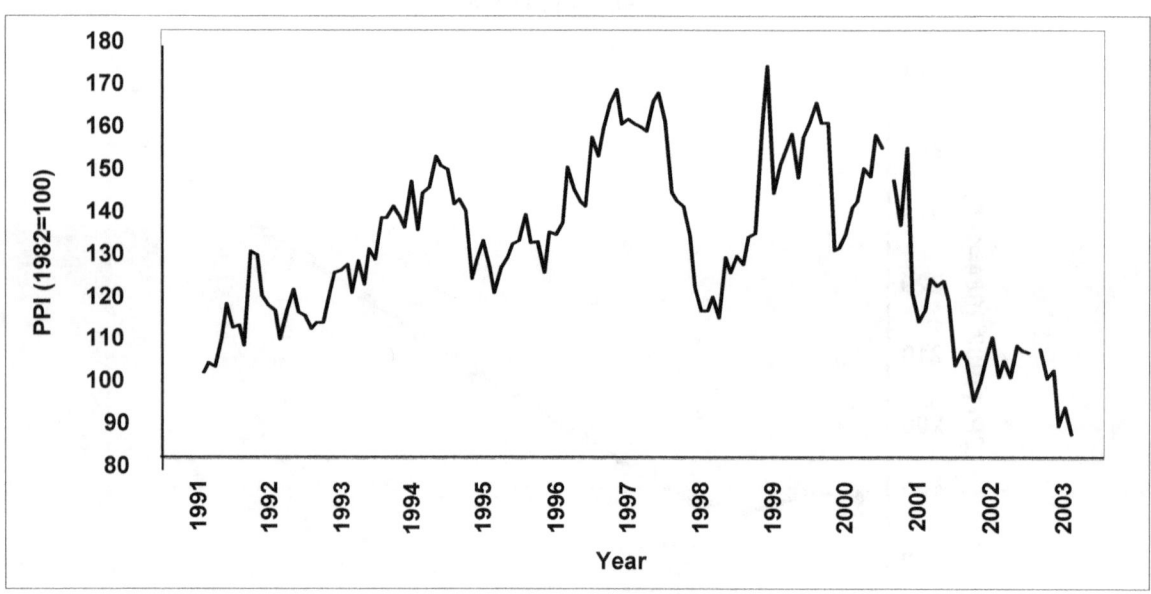

Source: BLS, 2004b.

Figure 3-6
Food Size Trout-Sales of Fish 12" or Longer
Annual U.S. Average Price per Pound

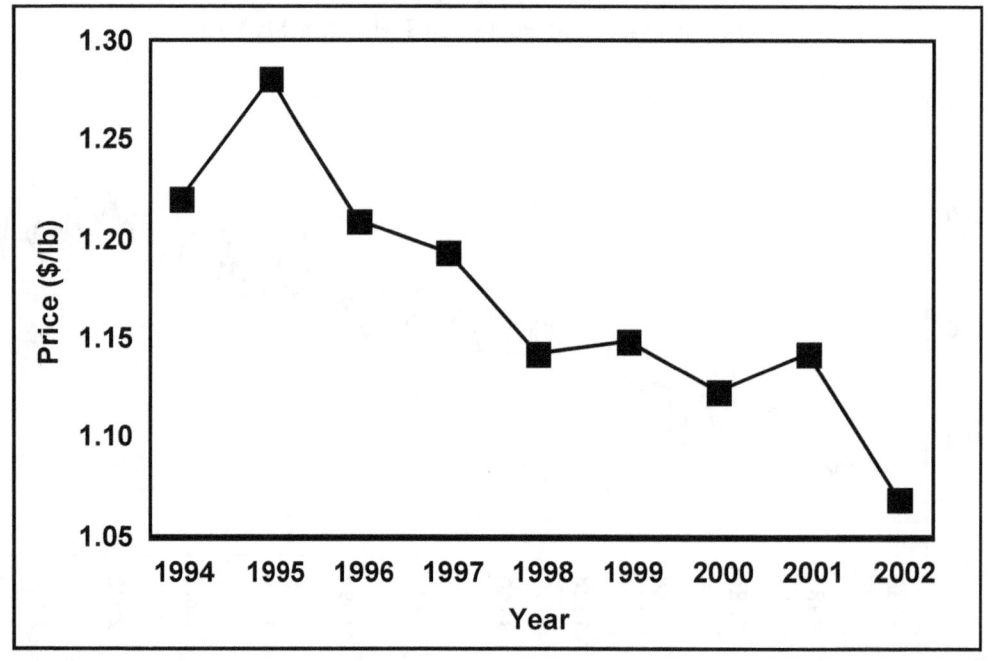

Source: USDA, 2003b.

Figure 3-7
Fulton Fish Market-Fresh Boned Idaho Trout
Monthly Price per Pound

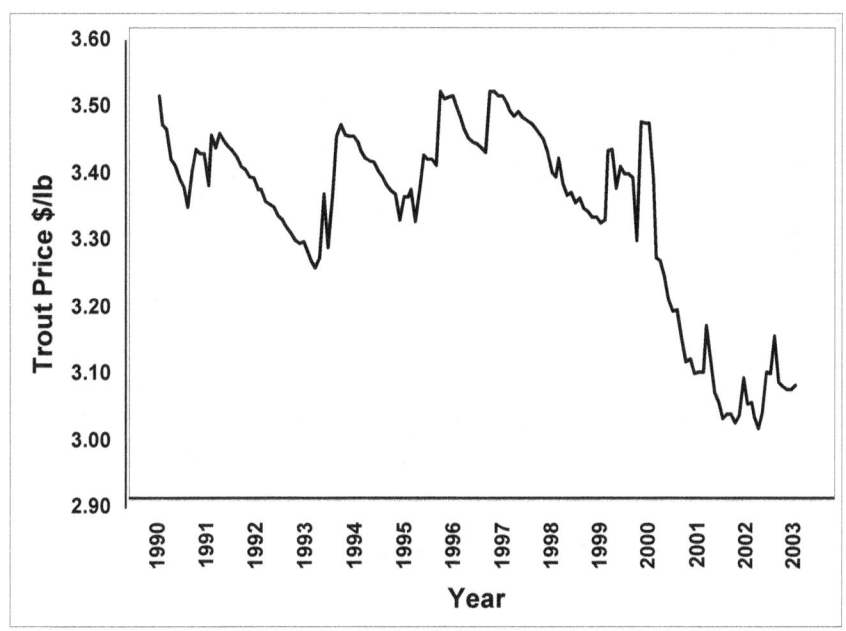

Source: NMFS, 2004.

Figure 3-8
Unprocessed and Packaged Fish-Monthly PPI

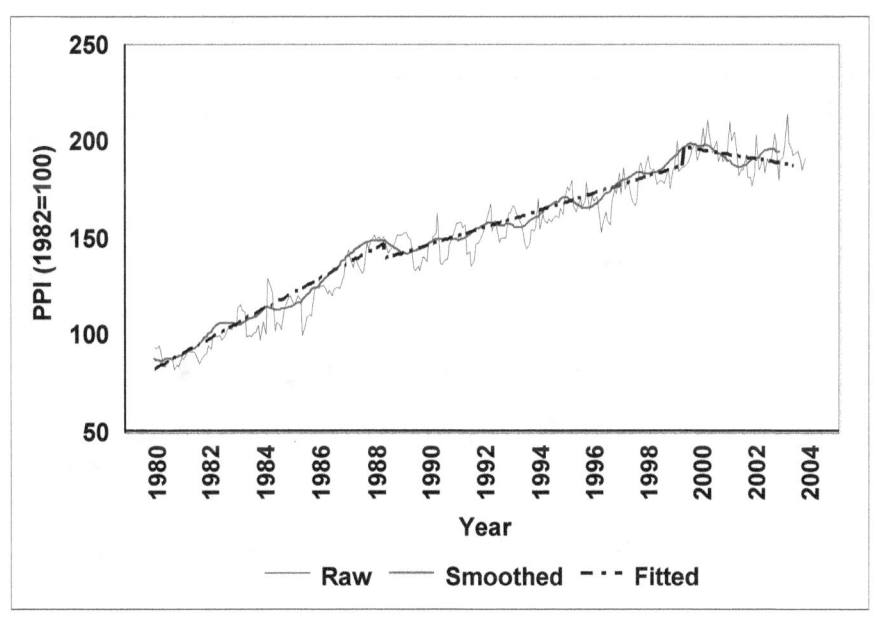

Source: BLS, 2004a, smoothed and fitted by EPA.

Figure 3-9
Shrimp Monthly PPI

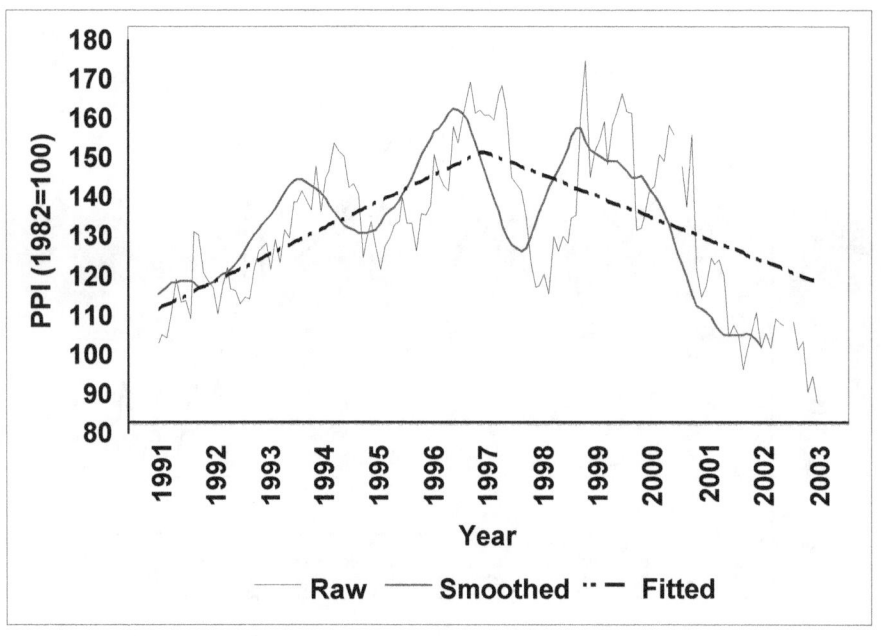

Source: BLS, 2004b, smoothed and fitted by EPA.

Figure 3-10
Food Size Trout-Sales of Fish 12" or Longer: Annual U.S. Average Price per Pound

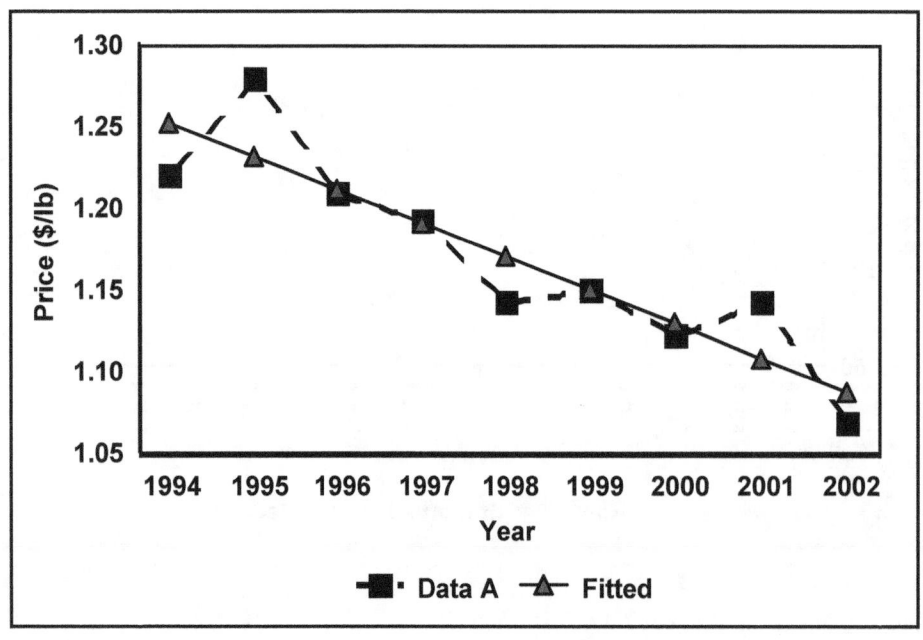

Source: USDA, 2003b, fitted by EPA.

Figure 3-11
Fulton Fish Market-Fresh Boned Idaho Trout: Monthly Price per Pound

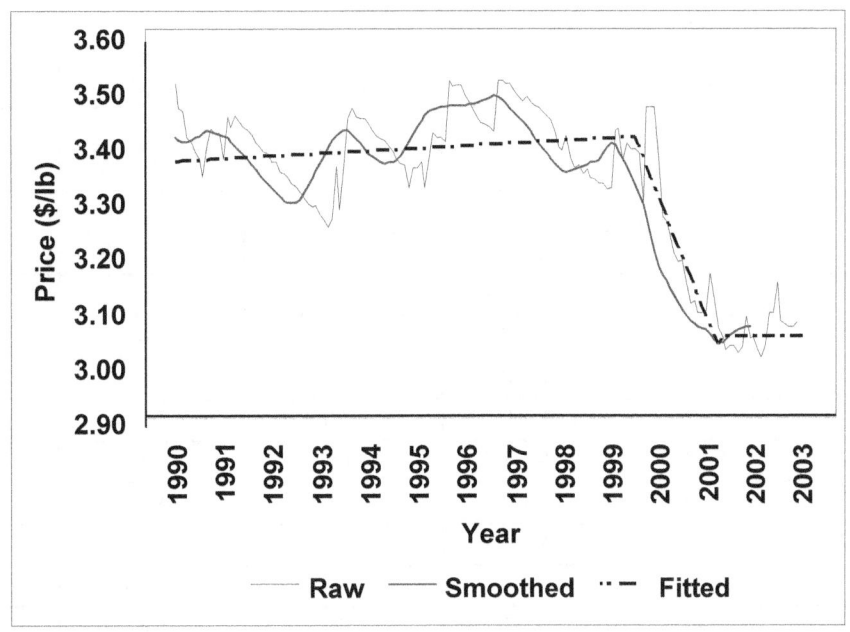

Source: NMFS, 2004, smoothed and fitted by EPA.

Figure 3-12
Forecasting Price Indices

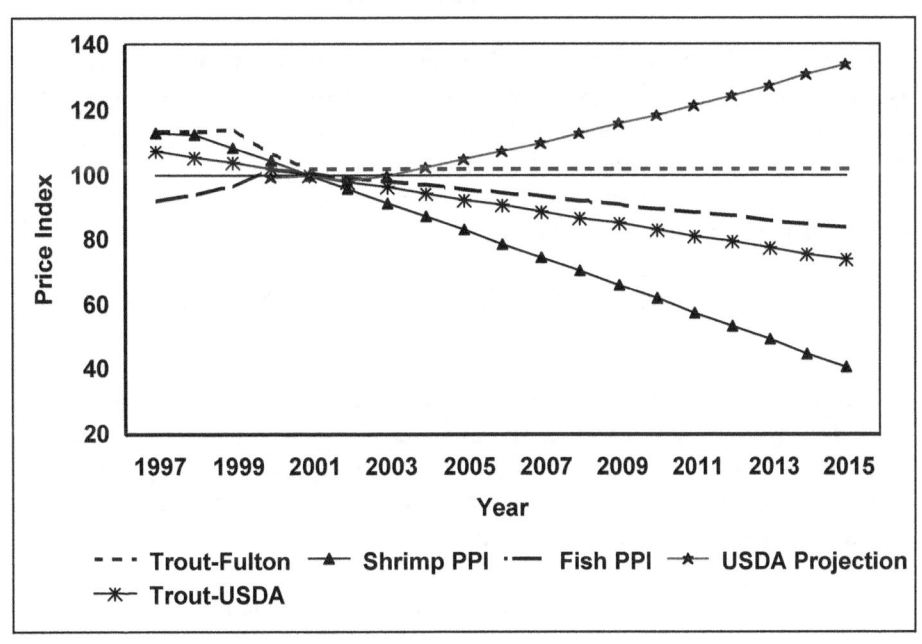

Source: EPA estimates.

Table 3-4
Forecasting Indices

| Year | Trout | | Shrimp | Fish | USDA |
	Fulton	USDA	PPI	PPI	Projection
2001	101.71	100.00	99.98	100.77	100.0
2002	101.71	98.15	95.77	99.58	98.4
2003	101.71	96.29	91.55	98.39	99.8
2004	101.71	94.44	87.32	97.20	101.8
2005	101.71	92.58	83.10	96.01	104.3
2006	101.72	90.73	78.88	94.82	107.0
2007	101.72	88.88	74.66	93.63	109.6
2008	101.72	87.02	70.43	92.43	112.3
2009	101.72	85.17	66.21	91.24	115.2
2010	101.72	83.32	61.99	90.05	118.1
2011	101.72	81.46	57.77	88.86	121.0
2012	101.72	79.61	53.54	87.67	124.0
2013	101.72	77.75	49.32	86.48	127.1
2014	101.72	75.90	45.10	85.29	130.3
2015	101.72	74.05	40.88	84.10	133.5

Sources: EPA estimates.

3.2.1.5 Pre-Regulatory Financial Conditions and Baseline Closures

EPA's closure analysis begins with an evaluation of the pre-regulatory financial condition of each in-scope facility. With three forecasting methods, there are three ways to evaluate a facility's future financial condition. If a facility's post-regulatory value is negative, the facility is flagged as a failure and assigned a score of "1" for that forecasting method. A facility, then, may have a score ranging from "0" to "3", failing under none to all forecasting methods. Several conditions may lead to a facility having a score of "2" or "3" under pre-regulatory conditions:

- The company does not record sufficient information at the facility-level for the closure analysis to be performed.

- The company does not assign costs and revenues that reflect the true financial health of the facility. Two important examples are cost centers and captive facilities, which exist primarily to serve other facilities under the same ownership. Captive facilities may show revenues, but the revenues are set approximately equal to the costs of the operation. (Cost centers have no revenues assigned to them).

- The facility appears to be in financial trouble prior to the implementation of the rule.

Under the first two conditions, the impacts analysis defaults to the company level because that is the decision-making level. For example, earnings data are held at the company level, not the facility level or the company has intentionally established facilities that will not show a profit but exist to serve the larger organization (e.g., cost centers or "captive" facilities). In either case, EPA does not have sufficient information to evaluate impacts at the facility level as a result of the rule.

The third condition identifies a facility with complete facility-level financial information and no confounding factors (i.e., it is not a captive or start-up facility) to obscure the financial condition of the facility. If the facility is unprofitable prior to the regulation, the company may decide to close the facility. This is likely to occur before the implementation of the rule to avoid additional investments in an unprofitable facility. The projected closure of a facility that is unprofitable prior to incurring costs associated with a regulatory action should not be attributed to the regulation. For the purposes of this analysis, EPA considers such facilities to be "baseline closures" and are not analyzed further.

This approach is consistent with established EPA practice. In the proposed rule and Notice of Data Availability, EPA characterizes baseline conditions using existing compliance levels and treatment in place (USEPA, 2002c and 2003). This approach is consistent with past effluent guidelines and EPA's Guidelines for Preparing Economic Analyses (USEPA, 2000a, Section 5.3.2) and Office of Management and Budget (OMB) guidelines. OMB guidelines state that "... the baseline should be the best assessment of the way the world would look absent the regulation ...It maybe reasonable to forecast that the world absent the regulation will resemble the present." (OMB, 2003). This means that if a facility already has option components in place by 2001 (the most recent year in the detailed questionnaire), EPA does not assign costs for those components to the facility. For example, if a facility has primary settling (a component that occurs in all options under consideration), it would not be expected to incur the costs for primary settling.

Similarly, EPA guidance indicates that facilities are not financially viable prior to the regulation, the closure of these facilities should not be attributed to the rule or meeting water quality criteria (USEPA, 2000a, p. 154 and EPA, 1995). These facilities are considered "baseline closures" (see Section 3.2.1.5 of this report). Costs for these facilities are not included in the cost of the rule (because the facilities are likely to close before the costs must be implemented); similarly, the pollutant removals associated with incremental pollution control for these facilities are not included in the cost-reasonableness or cost-effectiveness analyses.

Although the forecasting methods might be generalized as pessimistic, average, and optimistic, the pre- (and post-) regulatory status of a facility does not rest entirely on the results of the average forecast. Two of the forecasting methods start with 2001 data. If this was a good year, it is possible for both the pessimistic and optimistic forecasts to result in long-term positive earnings. If the same facility showed losses for 2000 and 1999, it is possible for the average forecast to result in long-term negative earnings. The reverse also holds. Of the 101 commercial facilities in this final regulation, the average forecast method did not coincide with the baseline status in 13 cases (i.e., the average forecasting method did not determine the baseline status for all facilities).

EPA performed a preliminary investigation of the baseline conditions for the industry. Of the 101 commercial facilities estimated to be within the scope of the rule, 32 are projected to be baseline failures. The number of facilities that can be analyzed for impacts of the rule is 69 facilities and all of them incur costs under the rule. Table 3-5 summarizes the baseline financial condition of the regulated aquaculture facility. Because the final regulation does not place different requirements for different production levels, the counts are not subcategorized by production size.

Table 3-5
Number and Types of Facilities in EPA's Economic Analysis for the Final Regulation

Production System	Owner	Estimated Number of Facilities		
		In-Scope	Baseline Closures	In Analysis and Cost Totals [1]
Flow Through and Recirculating	Commercial	82	24	58
	Noncommercial	139	NA	139
Net Pen	Commercial	19	8	11
	Noncommercial	0	NA	0
Alaska	Noncommercial	2	0	2
Total	Commercial	101	32	69
	Noncommercial	141	NA	141

Totals may not sum due to rounding. Earnings measured by cash flow.
NA: not applicable.
[1] In analysis counts are calculated by taking the number of in-scope facilities then subtracting out baseline closures.

EPA conducted an additional assessment of its baseline closure analysis and resultant baseline failures. Of the 32 facilities, 4 could not report financial data to EPA because of a recent change in ownership. Of the remaining 28, 18 reported 3 years of negative earnings in the questionnaire and none appeared to be start-up operations. The remaining 10 facilities have at least one year of positive earnings. For these facilities 2001 was a negative year. This is consistent with industry comments received by EPA.

EPA's closure analysis is based on cash flow as a measure of earnings and the three forecasting methods described above. Appendix A provides additional detailed discussions of several financial parameters used (or not used) to calculate earnings. The EPA survey collected financial data as reported in tax forms, so the discussion of earnings follows business terminology rather than agricultural economics terminology.

EPA performed several sensitivity analyses on the forecasting method. The first analysis used net income as a measure of earnings. The difference between net income and cash flow is depreciation. Depreciation is a non-cash cost and is excluded as a cost from cash flow and included as a cost when calculating net income. When net income is the measure of earnings, the number of baseline failures increases to 43, i.e., an additional 10 facilities cannot be analyzed for impacts. The second analysis considered any non-zero closure score in the main analysis as a closure. Under this assumption 34 facilities are considered baseline closures. The third and fourth analyses move the starting year for the forecasts to 2000 and compare the number of baseline failures under cash flow and net income projections. The year 2000 was, in general, a more profitable year for the industry than 2001 (see Figures 3-5 through 3-11). As expected, the number of facilities projected to be unprofitable prior to the rule drops from 32 to 27 for cash flow. The 10 facilities identified earlier with at least one year of positive earnings, but negative earnings for 2001 remain open when 2000 is the start year for the forecasts. Similarly, the number of facilities projected to be unprofitable prior to the rule drops from 43 to 40 under net income.

3.2.1.6 Projecting Facility Closures under the Final Regulation

Closure is the most severe impact that can occur at the facility-level and represents a final, irreversible decision in the analysis. The decision to close a facility affects the business owner, its workers, communities, and stockholders. When considering whether to terminate a business, the business will likely investigate several business forecasts and several methods of valuing their assets. Not only all data, assumptions, and projections of future market behavior would be weighed in the corporate decision to close a facility, but also the uncertainties associated with the projections. When examining the results of several analyses, the results are likely to be mixed. Some indicators may be negative while others indicate that the facility can weather the current difficult situation. A decision to close a facility is likely to be made only when the weight of evidence indicates that this is the appropriate path for the company to take. Thus EPA uses more than one forecasting method in the closure analysis.

EPA's analysis approximates financial decision-making patterns when determining when a facility would close. A score of "1" (implying negative earnings under only one of three forecasting methods) may result from an unusual year of data. When the score is "2" or "3", however, EPA considers that the weight of the evidence indicates poor financial health. EPA believes that this scoring approach represents a reasonable and conservative method for projecting closures. That is, a facility must show long-term financial health (e.g., a score of "0" or "1") prior to the incurrence of incremental pollution control cost and long-term unprofitability (e.g., a score of "2" or "3") after the incurrence of those costs.[5]

Facility closure represents a final, irreversible decision in the analysis. EPA estimates direct impacts from facility closures as the loss of all employment, production, exports, and revenue associated with the facility. This is an upper bound analysis, i.e., illustrating the worst effects because it does not account for other facilities increasing production or hiring workers in response to the closure of the first facility. The losses are aggregated over all facilities to estimate the national direct effect of the regulation.

3.2.1.7 National (Direct and Indirect) and Community Impacts

Impacts on aquaculture facilities affected by this final regulation are considered direct effects. Impacts due to reductions in production and employment by facilities that close are considered indirect effects. Induced effects are overall changes in household and business spending due to direct and indirect effects. The U.S. Department of Commerce's Bureau of Economic Analysis (BEA) tracks these effects both nationally and regionally in large "input-output" tables, published as the Regional Input-Output Modeling System (RIMS II) multipliers (DOC, 1996 and 1997). EPA uses the multipliers for the RIMS II industry number 1.0302 (miscellaneous livestock) because it includes all of SIC 0273. For this

---

[5] EPA's approach of not analyzing facilities with negative net earnings under "2" or "3" of the forecasting methods before they incur pollution control costs (i.e., baseline closures) is consistent with EPA guidance (USEPA, 2000a, EPA 240-R-00-003, p. 154).

analysis, EPA calculates direct and indirect impacts with the national-level final-demand multipliers. "Final demand" refers to the value of the sale to the final consumer of the product, a measure of change in production in the target industry. "Total" means that the multiplier includes direct, indirect, and induced economic effects. Total final demand multipliers show the relationship between the change in final demand in the target industry and change in output, earnings, or jobs in the whole regional economy. "Output" refers to overall production. It is what is measured by the Gross Domestic Product.

EPA uses national final demand multipliers for output and employment because they include direct, indirect, and induced effects. For this analysis, the national-level final demand multipliers for miscellaneous livestock are output ($3.7163 per $1) and employment (45.2228) full-time equivalents per $1 million in output in 1992 dollars.[6] For example, for every $1 million in output lost due to the projected closure of a regulated facility, nearly $3.8 million in output and 45 jobs are lost nationwide. When a facility is projected to fail as a result of the rule, EPA estimates the loss in output associated with facility closure, converts the loss to 1992 dollars with the Producer Price Index (CEA, 2004), and then use the RIMS II multipliers to estimate national level impacts.

If a facility is projected to close, all employment at the facility is considered lost. EPA evaluates the community impacts of facility closure by examining the increase in 2001 unemployment rate for the county in which the facility is located (BLS, 2003c). The increase in unemployment is calculated by adding the facility's employment to the county unemployment numbers and then recalculating the unemployment rate.

3.2.2   Other Regulatory Impact Criteria

In addition to its closure analysis, EPA also conducts additional analyses to assess potential effects on existing businesses. This includes an analysis of additional moderate impacts using a sales test, an evaluation of financial health using an approach similar to that used by USDA, and an assessment of possible impacts on borrowing capacity.

3.2.2.1 Sales or Revenue Test

To assess whether there are additional moderate impacts to facilities, EPA uses a sales test to compare the pre-tax annualized cost of the final rule to the 2001 revenues reported for facilities that passed the baseline closure analysis. EPA considers that facilities show additional moderate impacts if they are not projected to close but incur compliance costs in excess of 5 percent of facility revenue.

The threshold values EPA uses for its sales test (3 percent, 5 percent, and 10 percent) are those the Agency has determined to be appropriate for this rulemaking and are consistent with threshold levels used by EPA to measure impacts of regulations for other point source dischargers. EPA has used 1 percent and 3 percent sales test benchmarks to screen for potential impacts in many small business analyses (e.g., USEPA, 2000b, 2000c, 1997, and 1994). These benchmarks are only screening tools, but do support EPA's contention that a sales test of less than 3 percent generally indicates minimal impact.

---

[6] Employment multipliers are based on 1992 data, hence the loss in output needs to be in 1992 dollars.

The 5 percent threshold value that EPA uses for its sales test indicating potential "moderate" impacts are those the Agency has determined to be appropriate for this rulemaking and are also consistent with threshold levels used by EPA to measure impacts of regulations for other point source dischargers. For the final effluent guideline for Concentrated Animal Feeding Operations (CAFO), EPA defined farm operations with sales tests exceeding 5 percent but less than 10 percent as likely to incur moderate impacts (assuming simultaneously, positive cash flow or net income and acceptable debt to asset ratios), and, correspondingly, EPA defined farm operations with sales tests exceeding 10 percent as likely to indicate 'stressful' impacts (i.e., vulnerable to closure). For more information, see Section 3.4.1.

3.2.2.2 Credit Test

An additional test that EPA performs is a credit test that calculates the ratio of the pre-tax annualized cost of an option and the after-tax Maximum Feasible Loan Payment (MFLP) (i.e., 80 percent of after-tax cash flow). EPA identified any company with a ratio exceeding 80 percent of MFLP as affected under this test (i.e., the test threshold is actually 64 percent of the after-tax cash flow). These assumptions lend conservatism to the credit test.

While the closure analysis is performed at the enterprise, facility, and company levels, the credit test is performed at the company level only because this is the level at which financial institutions make their determination. The credit test population is the count of companies identified in the detailed questionnaire. Because the sample was drawn on facility characteristics, the survey weights apply to the facilities but not the companies that own them.

Based on the several measures used by USDA, EPA developed a method to examine whether a bank would lend a farm/company the amount needed to cover the costs of incremental pollution control. Like the financial health test, the credit test is performed with company-level data.

USDA notes that "Lenders generally require that no more than 80 percent of a loan applicant's available income be used for repayment of principal and interest on loans." (USDA, 2000a, p. 19). EPA considers the income available for debt coverage as the after-tax cash flow.[7] EPA chose the after-tax cash flow for 2001 (typically, the worst year in the questionnaire data) for the farm or company as the starting point for the credit test. For sole proprietors, EPA collected data for aquatic animal production from Schedule F or Schedule C from the IRS tax forms submitted with a proprietor's Form 1040. EPA intentionally did not request information from the proprietor's Form 1040 (the Agency specifically excluded the collection of off-farm income data). EPA multiplied the after-tax cash flow by 80 percent to obtain the proxy for USDA's "maximum feasible loan payment." EPA then calculated the ratio of the pre-tax annualized cost for an option to the after-tax MFLP. A more accurate, but less conservative, comparison would be the ratio of the after-tax annualized cost to after-tax MFLP, however, we assume a bank would be conservative and compare the pre-tax cost to the MFLP. As an additional measure of conservatism, EPA identified any facility with a ratio exceeding 80 percent as an impact (i.e., the test threshold is actually 64 percent of after-tax cash flow).

---

[7]The USDA definition for income for debt coverage is (net farm income + off-farm income + depreciation + interest - estimated income tax payments - family living expenses). EPA does not include off-farm income (a plus) or estimated family expenses (a negative); these are considered to off-set each other. EPA did not add interest back into the calculation.

3.2.2.3 Financial Health

EPA also calculates impacts on financial health at the company level using USDA's 2 x 2 matrix (four-state) categorization of financial health based on a combination of net cash income and debt/asset ratio (i.e., favorable, marginal solvency, marginal income, and vulnerable). EPA considers any change in categorization an impact of the final regulation.

Like the credit test, the financial health test is performed at the company level only, because of the requirement for a complete balance sheet. USDA's Economic Research Service (ERS) developed a financial performance measure that evaluates the combines effects of profitability and solvency (USDA, 2000a). Profitability is measured by a positive or negative income while solvency is measured by a debt/asset ratio. For this analysis, EPA uses the debt/asset ratio of 40 percent as the divider (USDA, 2000a). This results in a four-part classification for farm financial health:

- ▪ Favorable: positive income and debt/asset no more than 40 percent
- ▪ Marginal income: negative income and debt/asset no more than 40 percent
- ▪ Marginal solvency: positive income and debt/asset more than 40 percent
- ▪ Vulnerable: negative income and debt/asset more than 40 percent

For consistency with the closure model, EPA considers a company to have positive income if the present value of forecasted after-tax cash flow is positive in two or three forecasting methods.[8]

3.2.2.4 Other Financial Data and Criteria

EPA also considered other financial data and methodological approaches. In both cases, however, these data and methodological approaches were either not compatible with EPA's data and methodological needs or were deemed to be inconsistent with EPA's established guidelines for conducting regulatory analyses, or where not completed in time for consideration by EPA for this final rulemaking (given the Agency's notice and comment requirements).

University of Rhode Island Benchmarks. This project attempts to collect financial benchmark information for the aquaculture industry by researchers at the University of Rhode Island (URI) (Duguay, 2003). Professor Robert Comerford of URI, one of the principal investigators on the project, provided EPA with a draft copy of the report (Comerford and Rice, 2004).

Because the geographic area of interest is the Northeast and Rhode Island in particular, the large majority of the returned questionnaires represent oyster farms. Although the work is an impressive start on collecting benchmark financial data for use by the financial community, the URI data primarily represent production systems outside the scope of the rule. EPA notes one consistency between the URI data and EPA's projections on baseline financial conditions. Specifically, EPA's data show a substantial portion of aquaculture facilities do not report profits. Comerford and Rice (2004) also report that six of 15 respondents reported a profit in 2002 (40 percent) while 69 of the 101 in-scope commercial facilities (68 percent) are projected to be profitable.

---

[8]The USDA approach uses a single year of data.

Engle et al. study of U.S. Trout Operations. Prior to EPA's proposed regulations in 2002, various land grant university researchers participating in the JSA/AETF[9] indicated that they would be providing the Agency with enterprise budgets that could be used in an economic analysis. In March, 2004, EPA received a draft manuscript of an economic analysis based on several enterprise budgets for review (Engle et al, 2004). This study, referred to here as the "Engle study," reports that many aquaculture facilities already operate under consecutive years of negative profits and that EPA's regulation would further result in negative net returns at many regulated facilities, driving operators out of business.

The approach adopted in the study reflects an average representative facility (as opposed to a facility-specific analysis). For the purposes of assessing baseline financial conditions, this representative model approach differs markedly from EPA's facility-level approach as noted below.

The Engle study does not remove operations that show negative returns pre-regulation. Traditional EPA practice is to remove "baseline closures" from its regulatory analyses.

The Engle study uses collected financial data from its survey of trout operators in North Carolina and Idaho, including data from operations with negative returns, to compile representative farm financial budgets. The analytical approach the Engle study uses to analyze these average model budgets consists of a representative farm approach. Instead, EPA uses a facility-based approach, using financial data collected from it's detailed survey to analyze facility-level impacts for each facility type reflected in EPA's detailed survey for this rule. The approach used in the Engle study shows poor financial conditions, on average, because the sample includes financial data from non-viable firms (e.g., it includes what EPA would consider to be "baseline closures" in the averaging process).

The Engle study estimates financial conditions at regulated operations that includes cost estimates for "unpaid" labor. EPA's analysis does not account for unpaid labor because the Agency's detailed survey indicates that few (3 of an estimated 101 in-scope commercial facilities) report unpaid labor costs. EPA conducts sensitivity analyses of its results to account for potential unpaid labor costs among the three facilities that report such costs and the results reveal no additional impacts due to the rule. Because the Engle study accounts for unpaid labor costs using an average representative farm model approach, EPA believes that the resultant baseline financial conditions are artificially low.

EPA believes its facility-level approach, using actual financial data as reported in its detailed survey of regulated facilities, provides a more realistic picture of the baseline financial conditions at these facilities.

The Engle study approach also differs markedly from EPA's in its accounting of expected compliance costs and its evaluation of regulatory impacts. Specifically, the Engle study does not account for operations that are already complying with the expected regulatory options; nor does it account for existing "treatment-in-place" of various technologies at these facilities (i.e., the Engle study accounts for total economic costs regardless of what's already at the operation). EPA's facility-level approach utilizes the Agency's detailed survey information to determine facilities that either do not discharge wastewater to waters of the U.S. (and so would not incur any costs under the final regulation) or have some "treatment-in-place." Therefore, EPA's analysis reflects the expected "incremental" costs of complying with the final regulation, accounting for what the regulated facility already has in-place. EPA believes this

---

[9] Joint Subcommittee on Aquaculture's Aquaculture Effluents Task Force (see Section 2.1 of this report).

facility-level approach provides a more representative picture of the regulatory impacts existing facilities will incur as a result of complying with the final regulation, and therefore a better representation of economic achievability.

More information is available in the rulemaking record (USEPA, 2004a; ERG, 2004b; Tetra Tech, Inc., 2004). EPA's Response to Comments document provides additional detail.

Other issues raised in the Engle study are as follows. In particular, the study notes the financial hardship faced by regulated facilities and that producers regularly remain in operation despite consecutive years of negative income (e.g., because of certain non-economic factors as to why producers remain in business, such as lifestyle choice, farm operation as also home, inter-generational transfer of farm, and off-farm income supplementing family income). The study implies that EPA's regulatory analysis should consider sunk costs and include an allowance for capital replacement (in addition to including a proxy cost to reflect unpaid family labor and management). EPA's regulatory analysis addresses these cost items as follows. More detailed information is available in Appendix A of this report and also in EPA's responses to public comment.

- EPA believes that sunk costs paid out of capital (as opposed to financing) already occurred and, therefore, are not incremental cash flows. They should not affect future investment or the economic viability of the existing firm. Therefore, EPA excludes this category of sunk costs from the closure analysis. Sunk costs that are financed have interest, and this interest is included in interest payments reported in the income statements.[10]

- EPA includes costs for capital replacement as they occur within the depreciation and interest payments reported on an income statement. When EPA uses its net income calculations, capital replacement costs (as approximated by financial depreciation, in addition to interest payments captured in cash flow) are considered in the closure analysis. Capital replacement costs that are capitalized and not expensed are reflected in the asset, debt, and equity components of the balance sheet as appropriate. Past capital replacement costs are represented in the farm financial health measures and credit tests that are based on balance sheet data. When estimating compliance costs, EPA includes replacement costs for pollution control capital. EPA's cost estimates include all capital expenditures (whether initial or replacement) that are projected to occur within the 10-year analytical time frame.

Finally, the Engle study also raises issues associated with aquaculture entity's difficulty obtaining access to credit and limits to borrowing capacity given high absolute debt levels at these facilities. Aquaculture producers are characterized by high fixed and capital costs, and lenders are typically reluctant to issue loans to implement control treatments (since these are generally non-productive). Also lenders may attach a "risk premium" to the loan for specialty crops that effectively raises the interest rate.

---

[10]Question C6 of the detailed questionnaire asked the respondent to identify total expenses and individual expense items, such as interest payments (mortgage and other interest payments). Interest payments reflecting sunk costs are therefore included in total expenses for the facility and are therefore included in EPA's cash flow and net income analyses. The logic check for the questionnaire accepted the responses as long as total expenses exceeded the sum of individual items. That is, even if a respondent did not break out individual cost items, all interest payments for the facility would be included in total expenses.

EPA notes that similar issues were raised on an Agency rulemaking on the final CAFO regulation. For that rule EPA received no recommendations on possible approaches to deal with these issues as part of its analysis. Similarly for this final regulation, EPA did not receive any recommendations on how to address high debt levels and access to capital constraints among regulated facilities. EPA received no comments about its tests of borrowing capacity or financial health used in the analysis.

3.3   NONCOMMERCIAL FACILITIES

EPA initiated its consideration of the economic impacts on noncommercial facilities by reviewing the decisionmaking process where public hatcheries have been closed (Section 3.3.1). Section 3.3.1.1 discusses the National Fish Hatchery System while Section 3.3.1.2 focuses on State hatcheries in Alaska, Oregon, and Washington. Section 3.3.2 examines public reactions to potential hatchery closures and the role of user fees in funding state hatcheries. Section 3.3.3 reviews the two tests developed for analyzing the potential economic impacts of the rule on noncommercial facilities. Section 3.3.4 reviews the analysis for Alaska nonprofit organizations.

3.3.1   Closures of Noncommercial Facilities

3.3.1.1 National Fish Hatchery System

In 1999, the Fish and Wildlife Service (FWS) asked the federally chartered Sport Fishing and Boating Partnership Council to undertake a review of the role and mission of the National Fish Hatchery System (NFHS). The special report, entitled *Saving a System in Peril*, was published in 2000 (SFBPC, 2000). The report noted that, at that time, the NFHS had a maintenance backlog of about $300 million and that one out of every four hatchery personnel positions were vacant. The report highlights NHFSs role in three areas. These include: (1) Federal mitigation obligations, (2) restoring and maintaining native fisheries, and (3) recovery of threatened and endangered aquatic species.

Other recommendations (mostly in SFBPC, 2000, Attachment 5) include the following. First, hatcheries not needed to meet current or redirected program needs should be considered candidates for closure or transfer to States. Second, hatcheries should be evaluated for consolidation without loss in quality, production, or genetic diversity. Third, States should assume full management and financial responsibility for stocking public inland waters within their boundaries. Fourth, FWS should recoup all fish production costs for mitigation projects from the party or parties responsible for the development project.

FWS is acting on the report recommendations. Some recent closures include two facilities in Washington State and one in Arizona. The Willard NFH, Washington hatchery is scheduled to close in the spring of 2005. Thomas (2004) reports that the hatchery is operated with Mitchell Act funds[11] and Congress cut 2004 Mitchell Act funds by $1.1 million. The cuts are split among FWS and the State fish agencies in Washington and Oregon. The Eagle Creek NFH, Washington facility is scheduled for a reduction in coho salmon production (Thomas, 2004). At the Willow Beach NFH, Arizona facility, weekly releases of rainbow trout are scheduled to stop after March 2003 while facility is closed for

---

[11]A 1938 law that addresses compensating for fish losses caused by Columbia River dams.

raceway rehabilitation. It is reported that the hatchery will re-open with a focus on an endangered species (razorback suckers) and not trout stocking for recreation (Kimak, 2002).

3.3.1.2 State Hatchery Closures and Potential Closures

For the final rule, EPA collected information on how U.S. Fish and Wildlife Service (FWS) and State agencies make decisions about operating or closing public hatcheries. Much of this consists of information of actual State hatchery closures and/or potential closures. In particular, EPA obtained information on hatchery closings in three States—Alaska, Oregon, and Washington.

- **Alaska.** Heard (2003) reports that 13 Alaskan hatcheries closed since the program's inception. The closures occurred from 1979 (Fire Lake and Starrigawan) through 1998 (Eklutna) The reasons for the closures include: disease or genetic concerns for protecting wild stocks, avoiding major disease consequences in hatcheries, other biological concerns in the hatchery, management concerns over mixed stock fisheries, and cost efficiencies or other economic issues. Heard documented these closures, in part, to rebut an opinion that most "hatcheries, once built, continue to operate indefinitely, regardless of whether they achieve objectives and reach performance goals (Hilborn, 1999)..." (cited in Heard, 2003).

- **Oregon.** In September 2002, the Oregon Department of Fish and Wildlife announced plans to close four hatcheries: Cedar Creek, Elk River, Salmon River, and Trask River. Across-the-board budget cuts for all state agencies triggered the planned closures. The criteria for selecting the hatcheries to be closes included: source of funding (The four hatcheries are 100 percent General Fund supported), deferred maintenance costs at those facilities, costs of operations, and the cost of upgrading the facilities to meet new state and federal discharge permit requirements (ODFW, 2002a).

  The next day, the Oregon legislature approved a plan that shifted monies within different funds that—with conservative spending and good license sales—avoided the hatchery closures (ODFW, 2002b). All four hatcheries are listed on the ODFW web site with an "updated October 3, 2003" date (ODFW, 2003a).

  Prior to the September announcement, ODFW detailed the potential program cuts and held town meetings to discuss the public's reaction to the reduction in services (ODFW, 2002c and 2002d). The public meetings supported a fee increase but that the increase needs to support a specific purpose/service. The department announced the license and tag fees for the 2004 hunting and fishing seasons. The resident anglers have a $5.00 increase in the cost of an annual fishing license. The new cost is reported as $24.75, indicating a 20 percent increase (i.e., from $19.75 to $24.75). This is the first increase in fees in four years and the news release specifies that the fees will enable ODFW "to continue to support the public's priorities: hatcheries..." (ODFW, 2003b)

  Potential hatchery closures are not a new topic in Oregon. In 2000, ODFW proposed to close the Nehalem hatchery. ODFW attempted to reduce the range of impacts to fisheries by choosing a hatchery that was less diverse and less complex than others

(Nandor, 2000). The Nehalem hatchery is still on the ODFW hatchery list (ODFW, 2003a).

- **Washington.** Washington hatcheries are undergoing major reorganization based on the recommendations of the Puget Sound and Coastal Washington Hatchery Reform Project (LLTK, 2002; HSRG, 2002; McClure, 2002). A goal of the project is to protect and encourage the Puget Sound and coastal salmon stocks listed as threatened under the Endangered Species Act. One finding is that closure of some hatcheries and reduced production at others may benefit the survival of both native and hatchery fish. Among the 2002 recommendations are included: closing the McAllister Creek hatchery on the Nisqually River due to poor fish survival, disease transfer issues, and water quality (HRSG, 2002). (McClure, 2002 notes that the MacAllister Creek hatchery was one of three already scheduled for closure due to state budget cuts.) Closure identified as part of the Governor's 2002 supplemental budget (MRSC, 2002). The 2002 recommendations also include discontinuing hatchery programs for certain salmonids at Dungeness, Garrison Springs, Fox Island, Minter Creek, and Tulip Bay (MRSC, 2002). Seven other facility closures or possible closures in Washington may include: Fox Island, Sol Soc, Coulter Creek, Hurd Creek, Issaquah, and Percival Cove hatcheries.

- **Fox Island** net pens closed July 2001 (HSRG, 2002).

- **Naselle** Hatchery. The Washington Department of Fish and Wildlife (WDFW) budget lists the reasons for the possible closure of this hatchery as (1) a history of operational inefficiencies, (2) does not support a unique brood stock for the recovery of threatened salmonid stocks, and (3) low fish utilization (WDFW, 2003). Apalategui, 2003 notes that the hatchery rearing ponds are settling, cracking, and clogging with sediment and that the fish collection system allows the mingling of wild and hatchery stock. Potential closure is also noted in the Governor's budget summary (WA, 2003; MRSC, 2002).

- **Sol Duc** Hatchery. Proposed for closure in WDFW, 2003 budget which states that this will impact commercial and recreational fisheries in the Strait of Juan de Fuca but to a lesser degree than the closure of other hatcheries or hatchery programs. Potential closure is also listed in MRSC, 2002 but not WA, 2003.

- **Coutler Creek** Hatchery. Listed as possible closure in WA, 2003. HRSG, 2002 recommends that the chinook salmon releases be discontinued and that it would review the hatchery's role in yearling coho releases later. The report also notes that the site meets NPDES discharge levels but does not have a pollution abatement pond. Not listed as possible closure in WDFW, 2003 or MRSC, 2002.

- **Hurd Creek** Hatchery. Listed as possible closure in WA, 2003 and Kamb, 2003. Not listed as possible closure in WDFW, 2003 or MRSC, 2002. HRSG, 2002 notes that it is a satellite hatchery supporting the Dungeness hatchery. Kamb, 2003 mentions that Hurd Creek's captive brood program was scheduled to end in June 2004 and that HRSG recommended that the program be replaced with alternative methods.

- **Issaquah** Salmon Hatchery. Originally scheduled for closure in the early 1990's, the public formed a community group (FISH: Friends of the Issaquah Salmon Hatchery) and expanded the mission to include education as well as raising fish. However, the educational program means that the hatchery is more expensive to operate and the hatchery is listed as a possible closure (Goodman, 1997 and Holt, 2003).

■▯   **Percival Cove** Hatchery. The Washington Department of Ecology notified WDFW that is would not renew a discharge permit for aquaculture operations after April 2002 when that year's crop was released. The reasoning is the nutrient-laden fish food contributes to the phosphorus impairment of Capitol Lake (Dodge, 2002). The site is not on the list of WDFW hatcheries currently available on the department's web site (WDFW, 2004).

3.3.1.3 Summary

Based on EPA's review of public hatchery closures, EPA is conducting its analysis under the following assumptions.

First, public hatcheries close and it is therefore prudent to examine impacts of additional costs on public or noncommercial facilities.

Second, the reasons for closure vary widely. It may result from cuts to specific sources of funds (e.g., Mitchell Act funds and the Willard National Fish Hatchery or General Funds for Oregon hatcheries.). It may result from refocusing a program's mission and goals (e.g., toward Endangered Species Act concerns and away from recreational uses in the Federal and Washington State hatchery systems). Closure may also result from water pollution from aquaculture activities (e.g., Percival Cove in Washington), but those costs are not the primary reason for closure.

Third, the Federal hatchery system does not have user fee income. The refocusing of the Federal program on mitigation, native species, and endangered species is consistent with other suggestions that states should assume full management and financial responsibilities for stocking public fishing waters (SFBPC, 2000, p. 44). Therefore, inclusion of these facilities in the user fee test might be somewhat misleading. This is why the results were presented in three categories: no user fee, needed increase above a threshold, and needed increase below a threshold. On the other hand, the recommendations include FWS recouping all production costs for mitigation projects from the party responsible for the development project (SFPBC, 2000, p. 20). Were mitigation the only goal for the Federal hatchery system, we could assume a 100 percent cost-pass-through of any added pollution control costs once the recommendation is implemented.

3.3.2   State Hatcheries and User Fees

The reaction of the Oregon public to the possible closure of some state hatcheries, particularly the willingness to pay increased fees to keep them open (see Section 3.4.1.2), led EPA to investigate user fees and recent increases seen in user fees. A web search indicated that user fee increases do not happen every year and so the Agency compiled case studies where states increased fishing license fees.

EPA found seven recent examples of increases in fishing license fees: Pennsylvania, Nevada, New York, Oregon, South Carolina, Texas, and Wisconsin. Following are details on these examples:

■☐   **Pennsylvania** compiled a history of user fees in response to a proposed increase (Table 3-6; PF&BC, 2003a). The gap between increases ranges from three to 20 years. The increases range from 20 percent to 50 percent. The PA Fish and Boat Commission proposed an increase from $16.25 to $20.00 for 2004, a 23 percent increase (PF&BC, 2003b). It is not clear whether this change will pass through the legislature.

■☐   **Nevada** proposed a $5 increase in resident fishing licenses to be collected during the 2004 fiscal year (Sun, 2003). The increase would raise the price from $20 to $25, or 25 percent. Not only did this price increase go through, but the fee was raised an additional 16 percent to $29 for the 2004-2005 season (Henderson, 2004).

■☐   **New York** enacted an increase from $14 to $19 (36 percent) for a resident fishing licence for the 2002-2003 season (NY, 2003).

■☐   **Oregon** raised its resident fishing fees by $5.00 per license in 2003. The new cost is reported as $24.75, indicating a 20 percent increase (i.e., from $19.75 to $24.75). This is the first increase in fees in four years and the news release specifies that the fees will enable ODFW "to continue to support the public's priorities: hatcheries..." (ODFW, 2003b).

■☐   **South Carolina** increased the price of a resident combination hunting and fishing license from $20 to $25 for the 2003-2004 season. This is a 25 percent increase (Charleston, 2003).

■☐   **Texas** increased resident fishing license fees from $19 to $23 in 2003 (Texas, 2003a and 2003b). In January 2004, Texas created a new freshwater fishing stamp with a $5 price, resulting in a freshwater fishing license cost of $28 for residents (Texas, 2004). Texas, then, instituted two price increases of 22 percent to 24 percent each within two years.

■☐   In **Wisconsin**, the governor proposed raising the annual cost of a resident fishing license from $14 to $20, but the legislature trimmed the new cost to $17 (Chaptman and Jones, 2003). In other words, the governor proposed a 43 percent increase but the legislature accepted only a 21 percent increase.

The conclusion EPA draws from this information is that discrete user fee increases in excess of 20 percent are common. Increases as high as 36 percent have occurred in recent years.

Table 3-6
History of Pennsylvania Fishing License Fees

Year	Years Between Changes	Resident Fishing License	Change (%)	Trout/Salmon Stamp	Total Resident Cost for All Fish[1]
1922		$1.00			$1.00
1928	6	$1.50	50%		$1.50
1948	20	$2.00	33%		$2.00
1954	6	$2.50	25%		$2.50
1957	3	$3.25	30%		$3.25
1964	7	$5.00	54%		$5.00
1974	10	$7.50	50%		$7.50
1979	5	$9.00	20%		$9.00
1983	4	$12.00	33%		$12.00
1991	9	$12.00		$5	$17.00
1996	5	$16.25	35%	$5	$21.25
2003?	7				

[1]Excluding agent fees.
Source: PB&FC, 2003a.

Table 3-7 shows a list of 2003 resident state fishing license fees. These data show what an increase of $3 to $5 per license (typical of the raises mentioned above) would look like as percentages of the resident license fee. On a national basis, these fee hikes range from about 20 percent to 35 percent.

Table 3-7
2003 Resident Fishing License Fees

State	Resident License Fee 2003	Percent Increase over 2003 Fee		
		$3.00	$4.00	$5.00
Alabama	$9.50	32%	42%	53%
Alaska	$15.00	20%	27%	33%
Arizona	$18.00	17%	22%	28%
Arkansas	$10.50	29%	38%	48%
California	$30.70	10%	13%	16%
Colorado	$20.25	15%	20%	25%

State	Resident License Fee 2003	Percent Increase over 2003 Fee		
		$3.00	$4.00	$5.00
Connecticut	$20.00	15%	20%	25%
Delaware	$8.50	35%	47%	59%
District of Columbia	$7.00	43%	57%	71%
Florida	$13.50	22%	30%	37%
Georgia	$9.00	33%	44%	56%
Hawaii	$5.00	60%	80%	100%
Idaho	$23.50	13%	17%	21%
Illinois	$13.00	23%	31%	38%
Indiana	$14.25	21%	28%	35%
Iowa	$11.00	27%	36%	45%
Kansas	$18.50	16%	22%	27%
Kentucky	$15.00	20%	27%	33%
Louisiana	$9.50	32%	42%	53%
Maine	$22.00	14%	18%	23%
Maryland	$10.50	29%	38%	48%
Massachusetts	$27.50	11%	15%	18%
Michigan	$14.00	21%	29%	36%
Minnesota	$18.00	17%	22%	28%
Mississippi	$9.00	33%	44%	56%
Missouri	$11.00	27%	36%	45%
Montana	$17.00	18%	24%	29%
Nebraska	$16.00	19%	25%	31%
Nevada	$21.00	14%	19%	24%
New Hampshire	$35.00	9%	11%	14%
New Jersey	$22.50	13%	18%	22%
New Mexico	$18.50	16%	22%	27%
New York	$19.00	16%	21%	26%
North Carolina	$15.00	20%	27%	33%
North Dakota	$11.00	27%	36%	45%
Ohio	$15.00	20%	27%	33%
Oklahoma	$12.50	24%	32%	40%
Oregon	$19.75	15%	20%	25%
Pennsylvania	$17.00	18%	24%	29%
Rhode Island	$18.00	17%	22%	28%

State	Resident License Fee 2003	Percent Increase over 2003 Fee		
		$3.00	$4.00	$5.00
South Carolina	$10.00	30%	40%	50%
South Dakota	$21.00	14%	19%	24%
Tennessee	$21.00	14%	19%	24%
Texas	$19.00	16%	21%	26%
Utah	$26.00	12%	15%	19%
Vermont	$20.00	15%	20%	25%
Virginia	$12.50	24%	32%	40%
Washington	$21.90	14%	18%	23%
West Virginia	$11.00	27%	36%	45%
Wisconsin	$14.00	21%	29%	36%
Wyoming	$16.00	19%	25%	31%
Average	$16.34	21%	28%	35%

Source: PF&BC, 2003c.

3.3.3   Economic Tests

On the basis of the information EPA collected on noncommercial facility closures and on State user fees (Sections 3.3.1 and 3.3.2), the Agency developed two tests for evaluating the impacts of increased pollution control costs on noncommercial facilities. These are the budget test and an analysis of potential user fee increases.

3.3.3.1 Budget Test

The budget test compares the pre-tax annualized costs to the operating budget for the facility. As part of EPA's quality control process, the Agency examined the costs for part-time labor, full-time labor, and management as reported in Part B of the questionnaire with the total operating budget reported in Part C. Seven facilities reported labor and management costs that exceeded the operating budgets reported in Part C. A comment included with one of the facility surveys noted that the operating budget value did not include full-time labor or management. Presumably, part-time labor is considered a variable operating cost. For these seven facilities, EPA added the full-time labor and management costs to the reported operating costs.

For the 2002 Proposal, EPA assumed three different threshold values to evaluate an implied revenue test for noncommercial facilities: 3 percent; 5 percent, and 10 percent (see USEPA, 2002c and 2002d). EPA also requested comment on its approach and also recommendations on how to evaluate regulatory impacts to noncommercial facilities, including comment on its an implied revenue test threshold assumptions (USEPA, 2002c and 2003). EPA received no comments on its an implied revenue test and thresholds.

For the purposes of this analysis, EPA assumes a 5 percent and 10 percent threshold value as an indicator of potential financial impacts at noncommercial facilities. Accordingly, costs For more information about justification for these levels, see Section 3.4.1 and 3.4.2. These facilities would be affected by the final regulation unless they are able to raise user fees to cover these costs. As a supplemental analysis, EPA's analysis also considers how many government facilities that fail a given threshold can recover increased costs through funds from user fees (Section 3.3.3.2).

The use of these benchmark values is consistent with threshold values established by EPA in previous regulations for other point source dischargers. For more information, see Section 3.4.2.

3.3.3.2 User Fee Analysis

As part of a supplemental analysis, EPA also examines the ability of State-owned hatcheries to recoup compliance costs through increases in funding derived solely from user fees. This analysis is based on EPA's examination of the ability of State-owned hatcheries to recoup compliance costs through increases in funding derived solely from user fees.

All States and the District of Columbia have fishing license fees for residents. The license fees are not raised every year even though costs increase through inflation. Instead, when fees are raised or a fish stamp instituted, the raise or new fee is usually a round number such as $3, $5, or $10. A $3 to $5 hike in State fishing license fees translates into an increase in fees of about 20 percent to 35 percent.

The basis for this analysis is as follows. Part C of the detailed survey asked the respondent for the portion of the budget due from user fees, such as angler licenses, commercial fishing licenses, car vanity plates, and special purpose stamps. EPA examined the number of facilities that could pass through increased costs to the public through increased user fees and, where user fees were already in place, the size of the increase needed to cover the incremental costs. If the facility reports no revenue from user fees it is classified as no increase possible. Based on information presented in Table 3-7, user fee increases between 20 and 35 percent are not uncommon when they occur. Although all States report having fishing license fees, if a state hatchery reports no funding from user fee sources, EPA considers that facility to be unable to recoup increased costs through increased funding from user fees.

EPA believes that State facilities that receive user fee funds are those that are heavily invested in stocking streams for recreational angling (i.e., user fees are used to supply the fish that users catch). The availability of user fees might demonstrate additional flexibility in meeting additional costs, such as facilities that are facing higher compliance costs; these might be given access to user fee funds. As such, access to user fees might indicate greater flexibility to absorb additional costs associated with EPA's final regulation. The examples presented in this report of increases in user fees that States might be willing to adopt (e.g., ranges from 20 percent to 35 percent increase in a given year for States for which we were able to obtain data since 1980) demonstrates that States do in fact have the capacity to seek out additional funds, part of which goes to fish rearing facilities. EPA concludes, therefore, that States have demonstrated capacity to plan for increased costs, including potential compliance costs.

3.3.4   Alaska Nonprofits

Alaskan facilities perform ocean ranching where salmon smolts are released to the ocean. The members of the non-profit corporation are allowed to harvest adult fish that return to that region. These are reported as operator revenue. In addition, nonprofit hatcheries may allow region permit holders to vote for a self-imposed "enhancement tax" on the value of fish caught in that region (i.e., by member and non-member fishermen). EPA analyzed the impact of potential costs on Alaska nonprofit facilities by comparing the pre-tax analyzed cost to reported salmon revenues for 2001 in Alaska (2002). That is, grants, enhancement tax revenue, and income from miscellaneous sources such as visitor centers are excluded from the comparison.

3.4   EPA DECISION MATRIX FOR ECONOMIC ACHIEVABILITY

In general, effluent limitations guidelines represent the best economically achievable performance of facilities in the industrial subcategory or category ("Best Available Technology Economically Achievable" or "BATEA").[12] The Clean Water Act establishes BAT as a principal national means of controlling the direct discharge of toxic and nonconventional pollutants. The factors considered in assessing BAT include the cost of achieving BAT effluent reductions, the age of equipment and facilities involved, the process employed, potential process changes, non-water quality environmental impacts including energy requirements, economic achievability, and such other factors as the Administrator deems appropriate. The Agency retains considerable discretion in assigning the weight to be accorded these factors. Generally, EPA determines economic achievability on the basis of total costs to the industry and the effect of compliance with BAT limitations on overall industry and subcategory financial conditions. As with BPT, where existing performance is uniformly inadequate, BAT may reflect a higher level of performance than is currently being achieved based on technology transferred from a different subcategory or category. BAT may be based upon process changes or internal controls, even when these technologies are not common industry practice.

EPA's assessment of economic achievability for the final Concentrated Aquatic Animal Production (CAAP) regulation is complicated by the division of impacts across public and private facilities; as such, a single measure of economic achievability is not feasible. EPA's decision process for this final regulation is described below.

3.4.1   Commercial Facility Impacts

The primary measure of achievability for regulated commercial facilities is EPA's facility closure analysis. Secondary measures of moderate economic impacts include the sales or revenue tests, assuming a 5 percent criterion for a ratio of annual compliance cost to annual revenue, along with other measures of financial health and borrowing capacity (see Section 3.2.2). EPA also considers accompanying indirect and induced impacts on regional and national output and employment, given the results of its facility closure analysis.

---

[12]   Sec. 304(b)(2) of the CWA

The 5 percent threshold value that EPA uses for its sales test indicating potential "moderate" impacts are those the Agency has determined to be appropriate for this rulemaking and are also consistent with threshold values established by EPA in previous regulations. Generally, EPA's analyses have assumed that sales tests less than 5 percent indicate compliance costs that are achievable (see, for example USEPA 1994). Other analyses have assumed the same threshold but have further assumed that ratio values in excess of 5 percent may constitute moderate impacts, taking into consideration other factors (USEPA 2000b, and 1997). Sales impacts were assessed separately from those impacts that may make a facility vulnerable to closure. For the final effluent guideline for Concentrated Animal Feeding Operations (CAFO), EPA defined farm operations with sales tests exceeding 5 percent but less than 10 percent as likely to incur moderate impacts (assuming simultaneously, positive cash flow or net income and acceptable debt to asset ratios), and, correspondingly, EPA defined farm operations with sales tests exceeding 10 percent as likely to indicate 'stressful' impacts (i.e., vulnerable to closure). For more information, see EPA's Economic Analysis supporting the final CAFO regulations; EPA, 2002a.

For the purposes of assessing economic achievability, EPA assumes that the costs of the rule are not passed on to consumers, see Section 3.6 for a more detailed discussion.

3.4.2   Noncommercial Facility Impacts

Measures of achievability for regulated public facilities are restricted to the ratios of annual compliance costs to annual operating budgets, given the limited financial data and information on how to evaluate public facilities. EPA modeled the budget test, in part, on the sales test and chose the same thresholds to represent moderate and adverse impacts. In a sales test, the denominator in the ratio is sales (i.e., cost plus profit). In a budget test, the denominator is cost. Hence, a budget test is likely to be more stringent than a comparable sales test because of the absence of profit in the denominator.

For the purposes of this analysis, EPA assumes that those facilities that face costs exceeding 10 percent of their budget would be adversely affected by the final regulation, unless they are able to raise user fees to cover these costs. Operations where costs exceed 5 percent are considered to experience moderate impacts. EPA believes the 5 percent threshold value is reasonable given that noncommercial facilities obtain their operating revenues through Federal and State budget processes, which tend to more predictable year-to-year. Noncommercial facilities are also less susceptible to variability in overall market conditions that affect commercial operations. The use of a 10 percent benchmark as indicating possible adverse affects is also consistent with that assumed by EPA in previous regulations for commercial facilities (e.g., final CAFO regulations, see EPA, 2002a).

3.5   BARRIER-TO-ENTRY FOR NEW OPERATIONS

New Source Performance Standards (NSPS)[13] reflect effluent reductions that are achievable based on the best available demonstrated control technology. New facilities have the opportunity to install the best and most efficient production processes and wastewater treatment technologies. As a result, NSPS should represent the most stringent controls attainable through the application of the best available demonstrated control technology for all pollutants (i.e., conventional, nonconventional, and priority

---

[13] Sec. 306 of the CWA.

pollutants). In establishing NSPS, EPA is directed to take into consideration the cost of achieving the effluent reduction, any non-water quality environmental impacts, and energy requirements.

Typically, EPA evaluates impacts on new source facilities by comparing the costs borne by new source facilities to those estimated for existing sources. Accordingly, if the expected cost to new sources is similar to or less than the expected cost borne by existing sources (and that cost is considered economically achievable for existing sources), EPA considers that the regulations for new sources do not impose requirements that might grant existing operators a cost advantage over new source operators and further determines that the NSPS is affordable and does not present a barrier to entry for new facilities. If the expected cost to new sources is much greater than the cost borne by existing sources, this could discourage the start-up of new businesses who might not be able to compete with existing lower cost producers. In general, the costs to new sources from NSPS requirements are lower than the costs for existing sources because new sources are able to apply control technologies more efficiently than existing sources, which may incur high retrofit costs. Not only will new sources be able to avoid the retrofit costs incurred by existing sources, new sources might also be able to avoid the other various control costs facing some existing producers through careful site selection. If the requirements promulgated in the final regulation do not give existing operators a cost advantage over new source operators, then EPA assumes new source performance standards do not present a barrier to entry for new facilities.

For this analysis, EPA examines whether new aquaculture facilities would face barriers to entry because of the incremental pollution control costs under the final regulation. A barrier to entry analysis addresses the question whether the costs of incremental pollution control would increase the initial investment to the point where the person would change his/her mind on whether to start an operation.

The analysis include all facilities within the scope of the rule including those that fail the baseline discounted cash flow analysis. That is, the barrier to entry analysis includes facilities deemed to be "baseline closures" in the discounted cash flow analysis. Whether incremental costs constitute a barrier to entry is a different question from whether an unprofitable operation should continue to operate. For example, a failing site might incur zero costs under an option and that datum point should be retained in the analysis.

First, EPA examined the proportion of commercial facilities that incur no costs under each option. See EPA's *Development Document* for cost information (USEPA, 2004). Second, EPA examined the proportion of commercial facilities with no land or capital costs under each option. These comparisons examine facility costs and the calculations are the weighted proportions. Third, EPA examined the company average ratio of land and capital costs to total assets. This comparison is calculated on company data because asset data were collected only at the company level. Facility weights cannot be used for the company analyses. In this case, EPA calculates the ratio for each company and then uses the average of the ratios.

3.6   MARKET IMPACTS

EPA was not able to conduct a market model analysis for this rule for the following reasons. First, it is difficult to model this market given the interaction between commercial and noncommercial operations. For example, trout are raised commercially, but also for restoration and recreation. Second, wild catch accounts for a large share of the market for some species. For example, Alaskan salmon is considered a wild catch. Third, USDA Aquaculture Census data indicate that there is a high degree of concentration of specific species, such as trout and some other food fish. Fourth, there is insufficient data and analytical information to conduct this analysis. Specifically, literature on estimated measures of elasticity of supply and demand is limited and exist for only a few species, such as catfish which are not covered by this regulation. Elasticity measures do not exist for most other fish species. Because EPA was not able to conduct a market model analysis for this rule, the Agency is not able to report quantitative estimates of changes in overall supply and demand for aquaculture products and changes in market prices.

In addition, despite the fact that EPA does not have access to a market model as part of its analysis, there are other indications that long-term shifts in supply associated with this rule are unlikely given the dynamics of the U.S. aquaculture market. Specifically, the U.S. faces significant foreign competition from net-exporting nations and internal competition from wild catch and recreational catch harvests, among other factors.

These factors support EPA's approach of assuming that aquaculture producers are unable to pass on increased costs associated with this final regulation. This section discusses EPA rational for assuming that aquacultural producers are unable to pass on increased costs associated with this final regulation, which further highlights the Agency's determination that long-term shifts in supply associated with this rule are unlikely. Section 3.6.1 discusses the role of U.S. aquaculture compared to the world market. Section 3.6.2 reviews the competition within the U.S. from wild harvests and recreational fishing. Section 3.6.3 discusses industry concentration and producer-processor relationships and Section 3.6.4 is a summary.

3.6.1   U.S. Aquaculture Compared to Other World Aquaculture Markets

To evaluate the potential for trade and U.S. market impacts due to the final regulation, EPA collected information on world aquaculture from two sources: United Nations Food and Agriculture Organization (FAO) and U.S. National Marine Fisheries Service (NMFS). The numbers vary between the reports but the overall feature—the relatively minor position of the U.S. within world aquaculture—is consistent. FAO reports that total aquaculture production (including aquatic plants) was 45.7 million metric tons by weight and $56.5 billion by value in 2000 (FAO, 2002). China accounted for more than 70 percent of the total by volume and about 50 percent of the total value of world aquaculture production. Other major world producers included India, Japan, Indonesia, Thailand, Thailand, Korea, and several other Southeast Asian countries (Table 3-8).

According to the NOAA's National Marine Fisheries Service, world aquaculture produced 37.9 million metric tons in 2001. Estimated U.S. aquaculture production for 2001 is reported as 0.37 million metric tons or about one percent of total world production (NMFS, 2003). Thus, U.S. production is a small share of world production. Based on other available information, the value of U.S. aquaculture production accounts for roughly 2 percent of the world total. This is based on the reported value of U.S. aquaculture production at almost $1 billion in 2001 (NMFS, 2003), as compared to total world production during that same year, estimated at $56.5 billion including aquatic plants (FAO, 2002). Given the relative size of the U.S. aquaculture market, EPA concludes that the U.S. is unlikely to have much influence on import prices and the United States, in general, is likely a price taker rather than a price setter for aquaculture products.

Table 3-8
Major Aquaculture Producer Countries in 2000

Country	Quantity (1000 metric tonnes)	Value ($Million, US)
China	32,444	$28,117
India	2,095	$2,166
Japan	1,292	$4,450
Philippines	1,044	$730
Indonesia	994	$2,268
Thailand	707	$2,431
Korea, Republic	698	$698
Bangladesh	657	$1,159
Vietnam	526	$1,096
Rest of World [1]	5,200	$13,400
Total	**45,700**	**$56,500**

Source: FAO, 2002.
[1] Rounded to the nearest hundred. Estimated by EPA.

Farmed fish and other species serve as an important source of food for domestic markets, but exports are also an importance source of foreign trade. The main traded products from aquaculture are shrimp and prawns, salmon, and molluscs (FAO, 2002). In some cases, countries that are not among the top-ranked aquaculture producers in terms of overall production are among the top-ranked countries in terms of trade, particularly for individual fish specie categories. FAO (2002) notes that trade in farmed salmon went from zero to about 1 million metric tons in two decades with the majority of production coming from countries with limited domestic markets such as Norway and Chile. A large share of fish production enters international marketing channels, with about 40 percent exported in 2000 (live weight equivalent) in various food and feed product forms (FAO, 2002).

Across all species of traded fresh and frozen fish and shellfish, data from USDA's ERS indicate that the U.S. is a net-importer of seafood products. Table 3-9 shows the value of U.S. imports and exports of selected seafood products (USDA, 2002, Table 8). In 2001, U.S. exports totaled $0.6 billion, consisting of primarily salmon (frozen Pacific and unspecified canned and prepared salmon). During the same year, U.S. imports totaled $4.8 billion. Shrimp imports (both fresh and frozen) account for more than 75 percent of all imports (Table 3-9).[14] By weight, imports account for more than 40 percent of total

---

[14] If farmed, shrimp generally are raised in ponds which are not within the scope of the rule.

annual supplies. The U.S. also exports more than two-thirds by weight of its annual production, although mostly in frozen or processed form (Table 3-9). Historical data from USDA show that the gap between imports and exports has continued to widen during the 1990s, as the rate of increase in U.S. imports outpaced growth in U.S. exports.

Table 3-9
2001 Imports and Exports of Selected Seafood Products ($1000)

Product	Imports	Exports	Net
Shrimp, frozen	2,957,944	54,553	2,903,391
Atlantic salmon, fresh	685,289	37,945	647,344
Shrimp, fresh & prepared	678,853	51,481	627,372
Tilapia	127,797	0	127,797
Atlantic salmon, frozen	87,483	139	87,344
Mussels	43,610	1,595	42,015
Ornamental Fish	40,863	6,914	33,949
Oysters	36,914	8,238	28,676
Trout, fresh & frozen	11,507	1,577	9,930
Pacific salmon, fresh	30,462	22,166	8,296
Clams	8,296	6,593	1,703
Trout, live	99	271	(172)
Canned & prepared salmon	36,199	167,825	(131,626)
Pacific salmon, frozen	14,940	236,604	(221,664)
Total	**4,760,256**	**595,901**	**4,164,355**

Source: USDA, 2002.

One of the main reason the U.S. is not a major exporter of seafood products, as well as other types of agricultural products, is attributable in part to the presence of a large domestic market for these products. In the case of aquaculture, for example, although the U.S. exports about 2 million pounds of trout per year, this compares to total U.S. utilization of trout of roughly 100 million pounds annually, valued at about $76 million in 2001 (USDA, 2002). USDA data on U.S. aquaculture production of fresh and frozen trout show that U.S. imported 4.3 million pounds and exported 2.0 million pounds in 2003, consistent with the broader trends across the industry. For live trout, however, the U.S. currently reports a small positive net trade balance.

3.6.2   Intra-national Competition from Wild and Noncommercial or Public Sources

In addition to competition from foreign production, U.S. aquaculturists must also compete internally against the wild seafood harvest and production from noncommercial or public sources.

Production from wild seafood harvest greatly exceed that of farm-produced seafood products. The NMFS term for quantities of fish, shellfish, and other aquatic plants and animals brought ashore and sold is "landings." U.S. aquaculture's 2001 production was about 0.9 billion pounds (NMFS, 2003). In contrast, U.S. domestic landings for 2001 totaled 9.5 billion pounds. In terms of weight, wild catch is 11 to 12 times larger than domestic production. See Table 3-10 for a comparison by select species. In terms

Table 3-10
Sources and Uses of Aquaculture Species in the United States, 1998

Species	Units	Aquaculture		Wild Catch [1]	Net Imports	Total Use
		Total to Recreation, Restoration	Total to Food/ End use			
Catfish	(1,000 lbs)	10,175	563,934	11,590	1,100	586,799
		2%	96%	2%	0%	100%
Trout	(1,000 lbs)	46,341	47,422	789[1]	4,217	98,769
		47%	48%	1%	4%	100%
Salmon	(1,000 lbs)	291,147	107,160	644,434	42,331	1,085,072
		27%	10%	59%	4%	100%
Tilapia	(1,000 lbs)	0	11,571	0	60,911	72,482
		0%	16%	0%	84%	100%
Hybrid Striped Bass	(1,000 lbs)	612	8,407	6,715	1,927	17,661
		3%	48%	38%	11%	100%
Ornamentals	($1,000)	414	68,568	0	34,563	103,545
		0%	66%	0%	33%	100%
Baitfish	($1,000)	1,537	35,945	0[1]	0	37,482
		4%	96%	0%	0%	100%
Crawfish	(1,000 lbs)	35	17,426	22,226	4,387	44,074
		0%	39.5%	50.4%	10.0%	100%
Shrimp	(1,000 lbs)	8	4,209	277,757	670,212	952,186
		0%	0%	29%	70%	100%
Crab	($1,000)	21	10,276	473,378	295,518	779,193
		0%	1%	61%	38%	100%
Clam	($1,000)	50	50,026	135,237	31,164	216,477
		0%	23%	62%	14%	100%
Mussel	($1,000)	3	3,177	1,604	29,855	34,639
		0%	9%	5%	86%	100%
Oyster	($1,000)	27	26,985	88,627	29,785	145,424
		0%	19%	61%	20%	100%

Source: USDA, 2000b; USDA, 2000c; NMFS, 1998; and NMFS 1999.
[1] Figures shown for wild catch are from NMFS, 1999. Much of the trout and all of the baitfish wild catch is not reported to NMFS. Wild catch will be a substantial factor in both these markets.

of value, in 1998, U.S. aquaculture accounted for $0.9 billion while domestic landings accounted for $3.3 billion. Based on these data, aquaculture represents about 10 percent of the weight and about 30 percent by value compared to wild harvest.

Production from noncommercial or public facilities that primarily raise fish for ecological restoration, or recreation also account for a large share of total U.S. aquaculture production (depending on the species). Many of these fish are grown in government fish hatcheries; others are sold to government entities by commercial growers for stocking. Production decisions for these recreationally oriented growers are not governed by the same types of market forces that influence commercial decision-makers. Much of this production is financed by fishing license fees and other taxes. The ultimate consumers are anglers and those who value a natural environment. They do not make consumption decisions based on the price of stocking fish. Hence there is no market relationship, in the traditional sense, for these fish.

Table 3-10 summarizes the uses of aquaculture products and their sources for 1998 combining information from USDA's *1998 Census of Aquaculture* and National Marine Fisheries Service (NMFS) documents.[15] For example, almost half the trout and three-quarters of the salmon raised in U.S. aquaculture are used for ecological restoration, fee-fishing, or recreation (Table 3-10). Table 3-11 abstracts information from Table 3-10 to graphically illustrate the variety of market types among the aquaculture products.

3.6.3   Industry Concentration and Producer-Processor Relationships

The market structure for the private aquaculture industry is characterized by high facility concentration offset by competing sources and substitutes. USDA's *1998 Census of Aquaculture* data indicate a high degree of concentration at the facility level (USDA, 2000b). In the extreme cases, eight facilities in Texas produce 70 percent of the value of shrimp produced by aquaculture in the U.S. Three percent of the ornamental fish facilities (12 facilities) produce about 60 percent of the value of the industry. Table 3-12 summarizes the share of production from the top ten percent of facilities. Many of the aquaculture production industries are small and highly concentrated both in terms of the number of firms and geographic area (ornamentals, baitfish, salmon, and shrimp).

However, the existence of other sources of production, such as wild catch and imports, and close substitutes may limit the exercise of oligopoly power on the part of aquaculture producers. For salmon, shrimp, and most mollusks, the wild catch is greater than domestic aquacultural production. For baitfish, wild catch is not recorded in the fisheries statistics but is an important part of the market and always an option for anglers if farm-raised baitfish prices rise too high. Even when the wild product is only a close substitute for the farm-raised product, prices for the wild product will influence prices for the aquacultural product. If the wild catch products or imports are setting the price, it is unlikely that aquaculture producers could pass on increased production costs through to consumers and more of the costs of compliance (if not all) will need to be absorbed by the facility.

---

[15] Table 3-10 was assembled from three different sources so the data in each column may not be comparable to neighboring columns and adding them together may be incorrect. The purpose of the table, however, is to show rough scales of contributions of aquaculture (for recreation and food use), wild catch and imports to total U.S. supply for various species.

Table 3-11
Characteristics of Aquaculture Species Markets

Species	Aquaculture is largest source	Recreation is a large use	Imports...		Wild catch...	
			dominate domestic aquaculture	are a major component	dominates domestic aquaculture	is a major component
Catfish	X	-	-	-	-	-
Trout	X	X	-	-	-	[1]
Salmon	-	X	-	-	X	X
Tilapia	-	-	X	X	-	-
Hyb Striped Bass	X	-	-	X	-	X
Ornamentals	X	-	-	X	-	-
Baitfish	X	-	-	-	-	[1]
Crawfish	-	-	-	-	X	X
Shrimp	-	-	X	X	X	X
Crab	-	-	X	X	X	X
Clam	-	-	-	X	X	X
Mussel	-	-	X	X	-	-
Oyster	-	-	X	X	X	X

Source: Summarized from previous table.

[1] Much of the trout and all of the baitfish wild catch is not reported. Baitfish wild harvest was reported to be 50 percent of market at JSA Aquaculture Effluents Technical Workshop, 9/20/2000. Wild catch will be a substantial factor in both these markets.

Note: "Recreation is a large use" means ecological restoration, fee-fishing, recreational, and government use is greater than 20 percent of total use. "Dominates domestic aquaculture" means wild catch or net trade provides a greater proportion of total use than aquaculture. "Major component" means more than 10 percent of total use.

Table 3-12
Industry Concentration

| Species | Top 10 percent of farms | | Total Value ($1,000) |
	Number of Farms	Produce (Percent value)	
Catfish	137	65%	450,710
Trout	56	72%	72,473
Other Food Fish	44	85%	168,532
Ornamentals	35	75%	68,982
Baitfish	28	67%	37,482
Crustaceans	84	74%	36,318
Mollusks	54	79%	89,128

Source: USDA, 2000b.
Note: Production value categories added together to find top 10 percent.

Like wild catch, a high level of imports reduces the effect of changes in aquacultural production on the market. For tilapia, shrimp, and mussels, imports are a much larger share of the market than domestic aquaculture and undoubtedly have more influence on the market price. The situation for salmon is less straightforward, as the information presented in Tables 3-10 and 3-11 combine Pacific and Atlantic salmon. Also, the U.S. is net-exporter of processed salmon and frozen Pacific salmon, but a net-importer of fresh Atlantic salmon (Table 3-9). Atlantic salmon imports are twice total domestic salmon farm production. There is evidence that Atlantic and Coho salmon are substitutes in some situations (Clayton and Gordon, 1999). Whatever the precise relationships, trade flows have a large effect on the prices of many aquaculture products.

The largest segment of the U.S. aquaculture industry is catfish and is characterized by producers selling their goods to processors. USDA's *Aquaculture Outlook* and *Catfish Processing* reports the price paid by processors for farm-raised catfish and the average price received by processors for the final product (USDA, 2004b). However, catfish are raised mostly in ponds and not in the scope of the rule.

The salmon segment is marked by a limited number of companies. Some of these also own processing facilities as well, however, the pressure from imports will keep them from raising prices.

USDA (2004c) indicates that about 70 percent of food size trout (12 inches or longer) are sold to processors. In contrast, smaller-sized trout (between 6 to 12 inches) tend to be sold for to fee fishing operations (54 percent), the government (13.8 percent), and other producers (12 percent). Producers of food size trout are unlikely to have much market power because the majority of the fish are sold to processors. Producers of smaller trout have to compete with wild, recreational catch because most of their fish go to fee fishing operations. Finally, the trout segment is marked by many relatively small producers and thus trout producers have little ability to control prices.

3.6.4 Summary

In summary, EPA believes that its "no-cost-pass-through" assumption for the purposes of conducting its closure analysis is justified for the reasons discussed in this section. First, U.S. aquaculture production is small relative to the world market and the U.S. faces significant foreign competition from imports other lower-cost, net-importing nations. Second, U.S. aquaculture production also competes internally with production from both wild catch and recreational catch. Third, despite signs of concentration in these industries, the existence of other sources of production (e.g., wild catch, imports, and other close substitutes) may limit the ability of producers to control prices.

In addition, aquaculture operators are likely to have limited ability to pass on costs or negotiate higher product price, due in part to their position as suppliers of inputs to a complex chain of processors, wholesalers, and retailers (i.e., aquaculture operators have little influence over prices). Farmers are at the bottom of a long food marketing chain (including processors, wholesalers, retailers etc.) and cannot influence prices. Also, in part, this is attributable in part to imperfect market conditions characterized by "buyer" concentration (i.e., there are "few buyers" in the food processing and retail sectors relative to "many sellers" in the farm sector) and conditions of oligopsony/monopsony (Rogers and Sexton, 1994). Other factors include the competitive nature of agricultural production and the dynamics of the food marketing system. For more information, see EPA's Economic Analysis supporting the final CAFO regulations (USEPA, 2002a).

3.7 REFERENCES

Alaska. 2002. Alaska Division of Investments. Department of Community & Economic Developments. *Fisheries Enhancement: Revolving Loan Fund–Program Overview*. Third Edition. February.

Apalategui. 2003. Hatchery closure looms. *The Daily News*. Eric Apalategui. 17 November. Http://citizenreviewonline.org/nov_2003/hatchery.htm downloaded January 26 2004.

BLS (U.S. Department of Labor, Bureau of Labor Statistics). 2004a. Fish PPI, Producer Price Index - Unprocessed and packaged fish, not seasonally adjusted, Series ID:WPU0223, 1980:1- 2003:12, downloaded from http://data.bls.gov/labjava/outside.jsp?survey=wp, January 21.

BLS (U.S. Department of Labor, Bureau of Labor Statistics). 2004b. U.S. Department of Labor. Bureau of Labor Statistics. Shrimp PPI, Producer Price Index - Shrimp, not seasonally adjusted, Series ID:WPU02230501, 1991:12 - 2003:12 <http://data.bls.gov/labjava/outside.jsp?survey=wp> January 21.

BLS (U.S. Department of Labor, Bureau of Labor Statistics). 2003c. Local Area Unemployment Statistics. 2001 employment and unemployment data by state and county. <http://data.bls.gov/labjava/outside.jsp?survey=la>

Brealy, R.A. and S.C. Myers. 1996. *Principles of Corporate Finance*. 5th edition. The McGraw-Hill Companies, Inc. New York.

Brigham, E.F., and L.C. Gapenski. 1997. Financial Management Theory and Practice. Fort Worth, TX: The Dryden Press.

CCH (Commerce Clearing House, Inc.). 1999a. *2000 State Tax Handbook*. Chicago, IL.

CCH (Commerce Clearing House, Inc.). 1999b. *2000 U.S. Master Tax Guide*. Chicago, IL.

CCH (Commerce Clearing House, Inc.). 1995. Personal communication between Eastern Research Group, Inc., and CCH, Inc. (Commerce Clearing House, Inc.), to resolve discrepancies on tax rates for Missouri and Rhode Island. March 30.

CEA (Council of Economic Advisors). 2004. *Economic Report of the President*. Washington, DC. February.

Chaptman and Jones. 2003. Panel trims proposed increases in hunting, fishing license fees. *Milwaukee Journal Sentinel*. Dennis Chaptman and Meg Jones. 13 May. http://www.jsonline.com/news/state/may03/140616.asp?format=print downloaded 1/30/2004.

Charleston. 2003. Some license fees raised. *The Post and Courier*. 22 June. http://www.charleston.net/stories/062203/out_622_licensefees.shtml downloaded 1/30/2004.

Clayton, Patty L. and Daniel V. Gordon. 1999. From Atlantic to Pacific: Price Links in the US Wild and Farmed Salmon Market. *Aquaculture Economics and Management*, 3(2):93-104.

Comerford and Rice. 2004. *The University of Rhode Island Northeast Region Aquaculture Business 2003 Financial Benchmark Data*. Northeast Regional Aquaculture Center at the University of Massachusetts Dartmouth. Robert A. Comerford and Michael A. Rice. March draft.

DOC (U.S. Department of Commerce, Bureau of Economic Analysis). 1997. *Regional Multipliers: A User Handbook for the Regional Input-Output Modeling System (RIMS II)*. Washington, DC. March.

DOC (U.S. Department of Commerce, Bureau of Economic Analysis). 1996. Regional input-output modeling system (RIMS II). Total multipliers by industry for output, earnings, and employment. Washington, DC. Table A-24.

Dodge. 2002. State will close fish hatchery. *The Olympian*. John Dodge. February 2. http://news.theolympian.com/specialsections/Outdoors/20020202/9471.shtml downloaded January 26, 2004.

Duguay. 2003. Nicole Duguay. Professors develop first aquaculture database. *Pacer*. University of Rhode Island. April. <http://advance.uri.edu/pacer/april2–3/story22.htm> downloaded March 17, 2004.

Engle, C., S. Pomerlau, Gr. Fornstell, J.M.Honshaw, and D. Sloane. 2004. The Economic Impact of Proposed Effluent Treatment Options for Production of Trout *Orcorhynchus Mykr'ss* in Flow-Through Tank. Presented to the Joint Subcommittee for Aquaculture March.

ERG (Eastern Research Group). 2004a. Information Sources for Aquaculture Earnings Forecasts: USDA Trade Adjustment Program. Memorandum to Chris Miller and Renee Johnson, EPA from ERG. January 23.

ERG (Eastern Research Group). 2004b. Observations on Engle et alia, "The Economic Impact of Proposed Effluent Treatment Options for Production of Trout *Oncorhynchus mykiss* in Flow-through Tanks". Memoranda to Chris Miller and Renee Johnson, EPA. March 17 & 19.

FAO. 2002. Food and Agricultural Organization of the United Nations. The State of World Fisheries and Aquaculture: 2002. Rome.

FFSC (Farm Financial Standards Council). 1997. *Financial Standards for Agricultural Producers*. December.

Goodman. 1997. Hatchery models its success. *The Issaquah Press*. Stacy Goodman. http://www.issaquahhistory.org/press/articles/hatchery12-31-97.htm downloaded January 23, 2004.

Heard. 2003. Alaska salmon enhancement: a successful program for hatchery and wild stocks. in Nakamura, Y, J.P. McVey, K. Leber, C. Neidig, S. Fox, and K. Churchill (eds.). *Ecology of Aquaculture Species and Enhancement of Stocks*. Proceedings of the Thirtieth U.S.—Japan Meeting on Aquaculture. Sarasota, Florida, 3-4 December. UJNR Technical Report No. 30. Sarasota,FL:Mote Marine Laboratory. William R. Heard. Pages 149-169. http://www.lib.noaa.gov/japan/aquaculture/aquaculture_panel.htm downloaded January 27, 2004.

Henderson. 2004. Columnist Barb Henderson: increased license fees go to improvements. *Las Vegas Sun*. 23 January. http://www.lasvegassun.com/sunbin/stories/births/2004/jan/23/516226619.html downloaded January 30, 2004.

Hilborn, R. 1999. Confessions of a reformed hatchery basher. *Fisheries* 24(5):30-31. Cited in Heard, 2003.

Holt. 2003. Can hatchery fish survive the state's red ink? Issaquah operations staff brace for a 40% budget cut. *Seattle Post-Intelligencer*. 12 September. http://seattlepi.nwsource.com/local/139239_chatchery12.html downloaded January 23, 2004.

HSRG (Hatchery Scientific Review Group). 2002. Hatchery Reform Recommendations. February. http://www.lltk.org/hatcheryreform.html downloaded January 28, 2004.

IRS (Internal Revenue Service). 2003. How to Depreciate Property: for use in preparing 2003 returns. Publication 946. http://www.irs.gov/pub/irs-pdf/p946.pdf downloaded January 29, 2004.

IRS (Internal Revenue Service). 1999. The Complete Internal Revenue Code. Section 168(e)(D)(i) and Section 168(i)(13). July.

JSA (Joint Subcommittee on Aquaculture). 2003. Comments submitted on the proposed rule. DCN 70236.

Jarnagin, B.D. 1996. Financial Accounting Standards: Explanation and Analysis. 18th edition. Chicago, IL: Commerce Clearing House, Inc.

Kamb. 2003. Tribe fears recovery will be hampered under closure plan. *Seattle Post-Intelligencer.* 25 March. Page B1. http://seattlepi.nwsource.com/archives/2003/0303250197.asp downloaded January 28, 2004.

Kennedy, Peter. 1998. *A Guide to Econometrics.* The MIT Press. Cambridge, MA. 4th Edition.

Kimak. 2002. Outdoors: John Kimak. Future of trout production at Willow Beach hatchery unclear. *Las Vegas Review-Journal.* June 30. http://www.lvrj.com/cgi-bin/printable.cgi?/lvrj_home/2002/Jun-30-Sun-2002/living/19035335.html downloaded January 27, 2004.

LLTK (Long Live The Kings). 2002. Puget Sound and Coastal Washingon Hatchery Reform Project. http://www.lltk.org/pdf/hrbackground.pdf. downloaded January 27, 2004.

McClure. 2002. Hatchery reform takes big step forward. *Seattle Post-Intelligencer.* 20 February. http://seattlepi.nwsource.com/local/59008_hatch20.shtml downloaded January 23, 2004

MRSC (Municipal Research and Services Center of Washington). 2002. County related highlights of Governor's 2002 supplemental budget. http://www.mrsc.org/printfile.apsx?prntPath=%2ffocus%2fFile%2f0112cntybud.htm downloaded January 23, 2004.

Nandor. 2000. Frequently Asked Questions about the proposed Nehalem Hatchery closure. http://www.ifish.net/NFClosure.html. July. downloaded January 23, 2004.

NMFS (National Marine Fisheries Service). 2004. *Fisheries of the United States: 2002.* U.S. Department of Commerce. National Oceanic and Atmospheric Administration. Trout, Fulton Fish Market - *Fresh Prices at Fulton Fish Market*, Annual Summary, 1990-2002, Trout-Idaho-Boned, Monthly Average, 1990:1-2002:12. Downloaded from http://www.st.nmfs.gov/st1/market_news/index.html, January 20.

NMFS (National Marine Fisheries Service). 2003. *Fisheries of the United States: 2002.* U.S. Department of Commerce. National Oceanic and Atmospheric Administration. Silver Spring:MD. September.

NMFS (National Marine Fisheries Service). 1999. *Fisheries of the United States: 1998.* U.S. Department of Commerce. National Oceanic and Atmospheric Administration. Silver Spring:MD. Current Fishery Statistics No. 9800. July.

NMFS (National Marine Fisheries Service). 1998. *Imports and Exports of Fishery Products, Annual Summary, 1998.* U.S. Department of Commerce. National Oceanic and Atmospheric Administration. Silver Spring:MD.

NY (New York State Department of Environmental Conservation). 2003. Sporting license fee becomes law. 1 February. http://www.dec.state.ny.us/website/dfwmr/license/licfee02_03.html downloaded January 30, 2004.

ODFW (Oregon Department of Fish and Wildlife). 2003a. ODFW Fish Hatcheries. http://www.dfw.state.or.us/ODFWhtml/InfoCntrFish/odfw_hatcheries.htm updated 3 October 2003. downloaded January 23, 2004.

ODFW (Oregon Department of Fish and Wildlife). 2003b. 2004 fishing, hunting, and shellfish licenses go on sale Monday. Anglers and hunters will pay more for licenses and tags in 2004. News Release. November 26. http://www.dfw.state.or.us/public/NewsArc/2003News/November/112603news.htm downloaded January 27, 2004.

ODFW (Oregon Department of Fish and Wildlife). 2002a. ODFW announces plans for hatchery closures. News Release. September 17. http://www.dfw.state.or.us/public/NewsArc/2002News/September/091802bnews.htm downloaded January 23, 2004.

ODFW (Oregon Department of Fish and Wildlife). 2002b. Legislature's budget-balancing plan keeps hatcheries open, but more cuts scheduled if voters reject temporary income tax increase. News Release. 18 September. http://www.dfw.state.or.us/public/NewsArc/2002News/September/091902bnews.htm downloaded January 23, 2004.

ODFW (Oregon Department of Fish and Wildlife). 2002c. Summary of Town Hall Budget Meetings Held April 22-May 7, 2002. http://www.dfw.state.or.us/odfwhtml/200305_summary.html downloaded January 23, 2004.

ODFW (Oregon Department of Fish and Wildlife). 2002d. ODFW details fish and wildlife program cuts. News Release. March 13. http://www.dfw.state.or.us/public/NewsArc/2002News/September/091802bnews.htm downloaded January 23, 2004.

OMB (Office of Management and Budget). 2003. OMB Circular A-4, Regulatory Analysis. Informing Regulatory Decisions: 2003 Report to Congress on the Costs and Benefits of Federal Regulations and Impended Mandates on State, Local, and Tribal Entities. Appendix D and Appendix F. Washington, DC: Office of Management and Budget.

PF&BC (Pennsylvania Fish & Boat Commission). 2003a. Question of the Week: I heard the Fish and Boat Commission will be seeking an increase in fishing license fees soon. What's the history of fishing license fees in Pennsylvania? http://sites.state.pa.us/PAA_Exec/Fish_Boat/qfeehis.htm downloaded January 30, 2004.

PF&BC (Pennsylvania Fish & Boat Commission). 2003b. Commission welcomes license fee proposal from Sportsmen's Groups. Press Release. August 8. http://sites.state.pa.us/PA_Exec/Fish_Boat/newsreleases/2003/nwlicenseprop.htm downloaded January 30.

PF&BC (Pennsylvania Fish & Boat Commission). 2003c. Fishing license fees in other states. http://sites.state.pa.us/PAA_Exec/Fish_Boat/otherfishfee.htm downloaded January 30, 2004.

PL 108-27. 2003. Jobs and Growth Tax Relief Reconciliation Act of 2003. Public Law 108-07. May.

Rogers, Richard T. and Richard J. Sexton, "Assessing the Importance of Oligopsony Power in Agricultural Markets," *American Journal of Agricultural Economics*, Principal Paper, Vol. 76, No. 5, December 1994.

SFBPC (Sport Fishing and Boating Partnership Council). 2000. *Saving a System in Peril: A Special Report on the National Hatchery System.* http://sfbpc.fws.gov/Saving%20a%20System%20in%20Peril.pdf downloaded January 27, 2004.

Sun (Las Vegas Sun). 2003. Increases proposed for hunting, fishing licenses. January 31.

Tetra Tech, Inc. 2004. Analysis of *The Economic Impact of Proposed Effluent Treatment Options for Production of Trout* Oncorhynchus mykiss *in Flow-through Tanks*. Memorandum to Marta Jordan. March 25.

Texas (Texas Parks and Wildlife Department). 2004. Freshwater fishing stamp raises need to revamp licenses. News Release. 28 January. http://www.tpwd.state.tx.us/news/news/040128a.phtml?print=true downloaded January 30, 2004.

Texas (Texas Parks and Wildlife Department). 2003a. 2003-2004 hunting and fishing licenses go on sale Aug. 15. News Release. 4 August. http://www.twpd.state.tx.us/news/news/030804a.phtml?print=true downloaded February 2, 2004.

Texas (Texas Parks and Wildlife Department). 2003b. License Fees background. May 8. http://www.twpd.state.tx.us/involved/pubhear/proposals/fees-background.phtml downloaded February 2, 2004.

Thomas. 2004. Willard hatchery taking steps toward closure. *The Columbian.* 15 January. <http://www.columbian.com/01152004/sports/108906.html> downloaded January 26, 2004.

USDA (U.S. Department of Agriculture). 2004a. *USDA Agricultural Baseline Projections to 2013.* Prepared by the Interagency Agricultural Projections Committee. Staff Report WAOB-2004-1. February.

USDA (U.S. Department of Agriculture). 2004b. *Catfish Processing.* Report Aq 1 (2-02). National Agricultural Statistics Service. February 23.

USDA (U.S. Department of Agriculture). 2004c. *Production.* Aq2(2004). National Agricultural Statistics Service. February 27.

USDA (U.S. Department of Agriculture). 2003a. *USDA Agricultural Baseline Projections to 2012.* Prepared by the Interagency Agricultural Projections Committee. Staff Report WAOB-2003-1. February.

USDA (U.S. Department of Agriculture). 2003b. United States Department of Agriculture. *Trout Production*, Trout-Food Size, Sales of Fish 12" or longer, US Average Price per Pound, various dates, annual average 1994 - 2002. Downloaded from http://usda.mannlib.cornell.edu/reports/nassr/other/ztp-bb/, March 18.

USDA (U.S. Department of Agriculture). 2003c. Foreign Agricultural Service. 7 CFR Part 1580. Trade Adjustment Assistance for Farmers. *Federal Register* 68:50048-50053. August 20.

USDA (U.S. Department of Agriculture). 2003d. *Aquaculture Outlook.* LDP-AQS-17. Economic Research Service. March 14.

USDA (U.S. Department of Agriculture). 2002. *Aquaculture Outlook.* LDP-AQS-15. Economic Research Service. March 6.

USDA (U.S. Department of Agriculture). 2000a. *Agricultural Income and Finance Situation and Outlook.* AIS-75. Economic Research Service. September.

USDA (U.S. Department of Agriculture). 2000b. *1998 Census of Aquaculture.* Also cited as 1997 Census of Agriculture. Volume 3, Special Studies, Part 3. AC97-SP-3. National Agricultural Statistics Service. February.

USDA (U.S. Department of Agriculture). 2000c. *Aquaculture Outlook.* LDP-AQS-13. Economic Research Service. March 13

USEPA (United States Environmental Protection Agency). 2004a. *Development Document for the Final Effluent Limitations Guidelines and Standards for the Aquatic Animal Production Industry.* Washington, DC: U.S. Environmental Protection Agency, Office of Water.

USEPA (U.S. Environmental Protection Agency). 2004b. Memo to record documenting EPA's meeting to discuss an economic analysis of the impacts of EPA's CAAP regulation on the U.S. trout industry. May 7.

USEPA (U.S. Environmental Protection Agency). 2003. Effluent Limitations Guidelines and New Source Performance Standards for the Concentrated Aquatic Animal Production Point Source Category; Notice of Data Availability; Proposed Rule. 40 CFR Part 451. *Federal Register* 68:75068-75105. December 29.

USEPA (United States Environmental Protection Agency). 2002a. *Economic Analysis of the Final Revisions to the National Pollutant Discharge Elimination System Regulation and the Effluent Guidelines for Concentrated Animal Feeding Operations.* EPA-821-R-03-002. December.

USEPA (United States Environmental Protection Agency). 2002b. Detailed Questionnaire for the Aquatic Animal Production Industry. Washington, DC: OMB Control No. 2040-0240. Expiration Date November 30, 2004.

USEPA (U.S. Environmental Protection Agency). 2002c. Effluent Limitations Guidelines and New Source Performance Standards for the Concentrated Aquatic Animal Production Point Source Category; Proposed Rule. 40 CFR Part 451. *Federal Register* 67:57872. September 12.

USEPA (U.S. Environmental Protection Agency). 2002d. Economic and Environmental Impact Analysis of the Proposed Effluent Limitations Guidelines and Standards for the Aquatic Animal Production Industry. Washington, DC: U.S. Environmental Protection Agency, Office of Water. EPA-821-R-02-015. September.

USEPA (U.S. Environmental Protection Agency). 2000a. Guidelines for Preparing Economic Analyses. EPA-240-R-00-003. September.

USEPA (U.S. Environmental Protection Agency). 2000b. Economic Analysis of Final Effluent Limitations Guidelines and Standards for the Centralized Waste Treatment Industry. EPA-821-R-00-024.

USEPA (U.S. Environmental Protection Agency). 2000c. Economic Analysis of Final Effluent Limitations Guidelines and Standards for the Transportation Equipment Cleaning Industry Point Source Category. June. <http://www.epa.gov/ost/guide/teci>.

USEPA (U.S. Environmental Protection Agency). 1997. Economic Assessment for Proposed Pretreatment Standard for Existing and New Sources for the Industrial Laundries Point Source Category. EPA 821-R-97-008. November.

USEPA (U.S. Environmental Protection Agency). 1995. Interim Economic Guidance for Water Quality Standards. EPA-823-R-95-002. March.

USEPA (U.S. Environmental Protection Agency). 1994. Medical Waste Incinerators - Background Information for Proposed Standards and Guidelines: Analysis of Economic Impacts for New Sources. EPA 453/R-94-048a. July.

WA (State of Washington). 2003. Office of Financial Management. http://www.ofm.wa.gov/budget03/recsum/477rs.htm downloaded January 26.

WDFW (Washington Department of Fish and Wildlife). 2004. Hatchery Complexes and Facilities. http://www.wdfw.wa.gov/hat/facility.htm downloaded January 23, 2004.

WDFW (Washington Department of Fish and Wildlife). 2003. Proposed 2001-2003 Operating Budget Reductions: General Fund-State. Ongoing reductions. http://wdfw.wa.gov/pubaffrs/business/budgetreduction.htm downloaded January 26, 2004.

CHAPTER 4

REGULATORY OPTIONS:
DESCRIPTIONS AND COSTS

This chapter describes the final technology options that are the basis for the final rule and presents EPA's estimates of the national-level aggregate compliance costs to regulated facilities. Section 4.1 describes the technology options considered by EPA during the development of this rulemaking. Section 4.2 presents EPA's estimates of the number of affected facilities. Section 4.3 presents EPA's estimates of the expected pre-tax costs (2003 dollars) to these regulated facilities as a result of the final regulation. More detailed facility cost information is provided in EPA's *Development Document* supporting the final regulation (USEPA, 2004).

4.1   OPTION DESCRIPTION

4.1.1   Final Option

■   **Final Option** includes narrative standards for the control of solids based on implementation through BMPs addressing (1) feed management, (2) cleaning and maintenance, (3) storage of feed, drugs and pesticides to prevent spills, (4) record keeping on feed, cleaning, inspections, maintenance, repairs, and reporting requirements.

4.1.2   Options Discussed in 2003 Notice of Data Availability

Based on comments received on the proposed rule, detailed questionnaire data (which was not available at the time of proposal), and effluent monitoring (DMR) data received from EPA regional and State permitting authorities, EPA developed two additional options for consideration:

■   **Option A** includes (1) primary settling, (2) the requirement to develop and implement a BMP plan that minimizes both the discharge of drugs and chemicals and the possible escape of non-native species, and (3) the requirement for reporting Investigational New Animal Drugs (INADs) and extra-label use drugs as included in the proposed Option 2. The only difference between Option A and the proposed Option 2 is that Option A does not require the development and implementation of BMPs to address solids control.

■   **Option B** is similar to the proposed Option 3 in that it would require a greater degree of solids removal than achieved under Option A. However, Option B would offer facilities the choice to develop and implement a solids control BMP as included in Option 1 in lieu of installing secondary solids control technology, such as a second stage settling pond or a microscreen filter, and meeting numeric TSS limits. Facilities could still choose to install solids polishing technology and monitor TSS to achieve a numeric limit, but they could alternatively choose to instead implement solids control BMPs such as feed management.

4.1.3 Proposal Options

For the 2001 proposal (USEPA, 2002), EPA subcategorized the concentrated aquatic animal production (CAAP) facilities into flow-through, recirculating, and net pen production systems and considered three options for incremental pollution control:

■ **Option 1** for flow-through systems includes primary settling (e.g., quiescent zones and settling basins) and developing and implementing a BMP plan for solids control; for recirculating systems includes similar technologies/practices to those for flow-through systems; for net pens includes feed management and BMP plan development for solids control.

■ **Option 2** for all subcategories combining the Option 1 requirements with identifying and implementing BMPs to control discharges of drugs, chemicals, and non-native species; also includes a reporting requirement for the use of Investigational New Animal Drug (INAD) and extra-label use drugs.

■☐ **Option 3** combines Option 2 requirements with solids polishing (e.g., microscreen filtration) for flow-through and recirculating systems and active feed monitoring for net pens.

Table 4-1 identifies the components or technologies associated with each option for flow-through and recirculating systems. For net pen systems, Option B is the same as Option 3. EPA provided the public and the regulated community with the information about the additional options in its Notice of Data Availability (USEPA, 2003). This section summarizes EPA's estimated total regulatory costs for each of these options, including the technology option promulgated for the final regulation.

Table 4-1
Technologies or Practices by Option

Options	Technologies or Practices				
	Primary Settling	Solids Control BMPs	Drugs & Chemicals BMPs	Escape Prevention	Secondary Solids Removal
Final		✓	✓		
1	✓	✓			
2	✓	✓	✓	✓	
3	✓	✓	✓	✓	✓
A	✓		✓	✓	
B*	✓	✓	✓	✓	✓

* Option B would include primary settling, drugs and chemicals BMPs, escape prevention, and a choice between solids control BMPs or secondary solids removal technology.

4.2 TREATMENT IN PLACE AND BASELINE CONDITIONS AMONG COMMERCIAL OPERATIONS

The detailed questionnaire collected information at each facility as of 2001. EPA evaluates the treatment in place at each facility as of 2001. If a facility has an option component in place, EPA does not assign a cost for that component to the facility. The number of facilities that incur costs and are included in the impact analyses therefore varies by option. Table 4-2 summarizes the counts for the facilities that incur costs under the final regulation. Among the 101 commercial facilities, 32 are baseline closures. When net income is assumed as the basis for earnings, 43 facilities become baseline failures. That is, the use of cash flow for earnings results in a larger number of facilities in the cost and impact analysis for the industry. As discussed in Section 2.1 the count of noncommercial facilities includes Federal, State, Tribal, and Alaska nonprofit facilities.

Table 4-2
Estimated Number of Facilities With Production > 100,000 lbs/yr

Production System	Owner	Estimated Number of Facilities		
		In-Scope	Baseline Closures	In Analysis and Incur Costs[1]
Flow Through and Recirculating	Commercial	82	24	58
	Non-commercial	139	NA	139
Net Pen	Commercial	19	8	12
	Non-commercial	0	NA	0
Total	Commercial	101	32	69
	Non-commercial	141	NA	141

NA: not applicable.
[1]In-analysis counts are calculated by taking in-scope facilities then subtracting out baseline closures.

4.3 SUBCATEGORY AND INDUSTRY COSTS

The Notice presented costs for all facilities with flow-through, recirculating, or net pen systems that met the definition of a regulated concentrated aquatic animal production, i.e., the costs for facilities with 20,000 lbs/yr and greater production were included in the cost, nutrient cost-effectiveness, and cost-reasonableness analyses (USEPA, 2003). For promulgation, EPA is restricting the scope of the rule to facilities with greater than 100,000 lbs/yr of production. As a result, the detailed questionnaire data identified no academic/research operations within the scope of the rule.

The capital, one-time, and annual operating and maintenance costs are annualized using the approach described in Section 3.1. Annualized costs are by production system and owner. EPA estimates the annual incremental costs of compliance using the capital and recurring costs derived in the *Development Document* (USEPA, 2004). Annualized costs better describe the actual compliance costs that a regulated aquaculture facility would incur, allowing for the effects of interest, depreciation, and taxes. EPA uses these annualized costs to estimate the total annual compliance costs and to assess the economic impacts of the final requirements to each regulated operation. All costs presented in this section are converted from 2001 dollars to 2003 dollars using the Construction Cost Index (ENR, 2004).

Table 4-3 present EPA's estimated pre-tax and post-tax costs of the final regulation, respectively. The post-tax costs reflect the fact that a regulated operation would be able to depreciate or expense these costs, thereby generating a tax savings. Post-tax costs thus are the actual costs the regulated facility would face. Post-tax costs are also used to evaluate impacts on regulated facilities using a discounted cash flow analysis. Pre-tax costs reflect the estimated total social cost of the regulations, including lost tax revenue to governments. Pre-tax dollars are used when comparing estimated costs to monetized benefits that are estimated to accrue under the final regulations (see Sections 7 and 8 of this report).

Table 4-3
Pre-tax & Post-tax Annualized National Costs, Total by Subcategory and Option

Production System	Owner	Total Annualized Cost (Thousands, 2003 Dollars)[1]					
		Final	Option A	Option B	Option 1	Option 2	Option 3
Pre-Tax Annualized Cost							
Flow Through and Recirculating	Commercial	$256	$90	$258	$194	$251	$634
	Non-commercial	$1,149	$717	$1,382	$1,221	$1,384	$2,235
Net Pen	Commercial	$36	$0	$0	$0	$0	$0
	Non-commercial	$0	$0	$0	$0	$0	$0
Total		**$1,442**	**$807**	**$1,640**	**$1,415**	**$1,635**	**$2,869**
Post-Tax Annualized Cost							
Flow Through and Recirculating	Commercial	$202	$79	$203	$146	$197	$565
	Non-commercial	$1,149	$717	$1,382	$1,221	$1,384	$2,235
Net Pen	Commercial	$11	$0	$0	$0	$0	$0
	Non-commercial	$0	$0	$0	$0	$0	$0
Total		**$1,362**	**$796**	**$1,585**	**$1,367**	**$1,581**	**$2,800**

Note: Totals may not sum due to rounding.
Estimated by EPA.
[1] EPA converted costs from 2001 dollars to 2003 dollars using the Construction Cost Index (Engineering News Record, 2004). Costs are for facilities that are not baseline closures under a cash flow analysis.

For noncommercial facilities, the cost estimates are the same in both the pre-tax and post-tax tables since EPA assumes no tax savings for these facilities. EPA estimates national costs on the number of facilities expected to incur compliance costs if they exceed the production threshold in the final rule. That is, EPA includes all facilities that are not baseline closures and those for which EPA could not make a baseline closure determination (e.g., start-up operations or facilities with insufficient data) under the cash flow assumption.[16]

The estimated annualized costs for the final regulation is $1.4 million. Noncommercial facilities account for about 80 percent of the total cost of the rule. These estimated total costs reflect aggregate compliance costs incurred by facilities that produce more than of 100,000 lb/year and will be affected by today's final regulation.

For comparison, Table 4-3 also presents estimated costs across a range of technology options considered by EPA during the development of this rulemaking.

4.4   COST-REASONABLENESS

EPA performed an assessment of the total cost of the final rule relative to the expected effluent reductions. EPA based its "Cost Reasonableness" (CR) analysis on estimated costs, loadings, and removals. EPA estimates BOD and TSS removals for each facility for each option. Because BOD can be correlated with TSS, EPA selected the higher of the two values (not the sum) to avoid possible double-counting of removals. Option costs include costs for certain BMP components that are not part of the final rule address [need clarification]. That is, EPA's cost-reasonableness values are likely overstated.

The Cost Reasonableness for the Flow Through and Recirculating subcategory is $2.77. Cost-reasonableness is undefined for the Netpen subcategory because facilities in this groups has adequate treatment to achieve requirements for pollutants (i.e., no incremental removals are estimated). See EPA's Development Document (USEPA, 2004) and ERG, 2004 in the rulemaking record for additional details.

4.5   REFERENCES

ERG (Eastern Research Group). 2004. Aquaculture Cost-Reasonableness. Memorandum to Chris Miller , EPA. June 10.

ENR (Engineering News Record). 2004. Construction cost index history, 1908-2004. Engineering News Record. Downloaded April 1, 2004
http://enr.construction.com/features/conEco/costIndexes/constIndexHist.asp

USEPA (U.S. Environmental Protection Agency). 2004. Development Document for the Final Effluent Limitations Guidelines and Standards for the Aquatic Animal Production Industry. Washington, DC: U.S. Environmental Protection Agency, Office of Water.

---

[16] The number of baseline closures increases under net income analysis, implying that national costs decrease under EPA's net income analysis.

USEPA (U.S. Environmental Protection Agency). 2003. Effluent Limitations Guidelines and New Source Performance Standards for the Concentrated Aquatic Animal Production Point Source Category; Notice of Data Availability; Proposed Rule. 40 CFR Part 451. *Federal Register* 68:75068-75105. December 29.

USEPA (U.S. Environmental Protection Agency). 2002. Effluent Limitations Guidelines and New Source Performance Standards for the Concentrated Aquatic Animal Production Point Source Category; Proposed Rule. 40 CFR Part 451. *Federal Register* 67:57872-57928. September 12.

CHAPTER 5

ECONOMIC IMPACT RESULTS

This section presents the national-level aggregate compliance costs and economic impacts on regulated facilities under the final regulations.

Section 5.1 presents EPA's estimated impacts to existing sources to comply with the guidelines and standards established under the effluent limitations guidelines program, including Best Practicable Control Technology Currently Available (BPT), Best Available Technology Economically Achievable (BAT), Best Conventional Pollutant Control Technology (BCT), and New Source Performance Standards (NSPS).[1] Results are presented for both commercial and noncommercial facilities. This section also provides a brief comparison of EPA's economic impact analysis of the options considered for the proposed rule in 2002 and other technology options considered by EPA during the development of this rulemaking. More detailed analysis of these other technology options is provided in supporting documentation on the proposed regulation (see USEPA 2002a and 2002b) and in the Agency's Notice on the proposed rule (USEPA, 2003).

Section 5.2 examines the impact to new facilities on complying with the final effluent guideline requirements for New Source Performance Standards (NSPS) and presents EPA's barrier to entry analysis for new sources.

Finally, Section 5.3 presents EPA's assessment of the potential market-level analysis, including the effects of the regulation to U.S. trade, consumer markets, and community level impacts.

5.1 BEST AVAILABLE TECHNOLOGY FOR EXISTING SOURCES (BPT, BAT, AND BCT)

This section presents the results of EPA's analysis of the economic impacts on existing commercial and noncommercial operations. Table 5-1 shows the results of EPA's regulatory impact analysis for both commercial and noncommercial operations.

5.1.1 Commercial Facilities

There are 101 commercial facilities within the scope of the rule. To evaluate impacts to commercial facilities, EPA conducts a closure analysis that compares projected earnings with and without cost of compliance with the final regulation for the period 2005 to 2015. EPA's analysis examines possible closures at three different organization levels: enterprise, facility, and company; results for facilities are presented in Table 5-1. In addition to its closure analysis, EPA assesses other potential effects, considered as "moderate impacts", using a sales test, an evaluation of financial health using an approach similar to that used by USDA, and an assessment of possible impacts on borrowing

---

[1] Since EPA is not promulgating standards for indirect dischargers, the analysis does not include a discussion for Pretreatment Standards for Existing Sources or Pretreatment Standards for New Sources.

Table 5-1
Economic Effects: Existing Commercial & Noncommercial Operations

Threshold Test	Estimated Number of In-Scope Facilities	Final Option
Commercial Operations		
Closure Analysis[1]	**101**	**0**
Sales test >3%	101	4
Sales test >5%	101	4
Sales test >10%	101	0
Change in Financial Health	NA[2]	0
Credit test >80%	NA[2]	1
Noncommercial Facilities[5]		
Budget test >3% (all facilities)	**141**	**19**
State owned only (# with user fees)[4]	106	12 (8)
Federal owned only	33	7
Alaskan Non-Profit[3]	2	0
Budget test >5% (all facilities)	**141**	**12**
State owned only (# with user fees)[4]	106	8 (8)
Federal owned only	33	4
Alaskan Non-Profit[3]	2	0
Budget test >10% (all facilities)	**141**	**4**
State owned only (# with user fees)[4]	106	0 (0)
Federal owned only	33	4
Alaskan Non-Profit[3]	2	0

Source: Estimated by USEPA using results from facility-specific detailed questionnaire responses, see Chapter 3.

1) Closure analysis assumes discounted cash flow for earnings. A total of 32 facilities are projected to be baseline closures; these facilities cannot be attributed to this rule.

2) Analysis performed at the company level. EPA evaluated 34 unweighted companies representing the 101 weighted facilities from the detailed questionnaire. The statistical weights, however, are developed on the basis of facility characteristics and therefore cannot be used for estimating the number of companies.

3) Two Alaska non-profit organizations are within the scope of this rule, but did not receive a detailed survey. They were costed using screener survey data. Economic impacts were calculated using publically available information.

4) Some State-owned facilities reported that they relied, in part, on funds from State user fee operations. These numbers are reported in parenthesis and are included in the overall numbers as well.

5) EPA maintains that there is potential for Tribal facilities to be present within the population of noncommercial facilities affected by this rule, despite the absence of a line item for Tribal facilities above. EPA, recognizing that the mission of Tribal facilities may differ to some extent from the mission of State and Federally operated facilities, maintains that operating budgets, standardized for production level, are likely to be similar to those presented in Table IX-3 (approximately 3% and 9% respectively).

5.1.1.1 Closure Analyses

EPA projects no closures as a result of the rule for the 8 enterprises, 101 facilities, or 34 companies determined to be in-scope. Projections were based on cash flow as a measure of earnings and 2001 as the starting year for earnings forecasts (Table 5-1). Results for sensitivity analyses regarding assumptions used to assess closures are presented in Section 5.1.1.3. Note that all other analytical results (e.g., costs, cost reasonableness, benefits) reflect cash flow and negative earnings in less than 2 of 3 forecasts. Further information on the characteristics of companies, facilities, and enterprises determined to be in-scope of the rule are contained in Chapter 2. For the purposes of this analysis, EPA assumes these operations are not able to pass on the compliance costs due to the regulation. EPA's assumption of "no cost pass through" is a more conservative approach to evaluating economic achievability among regulated entities, see Section 3.6 of this report. (To evaluate market and trade level impacts, however, EPA assumes all costs are shifted onto the broader market level as a way of assessing the upper bound of potential effect, see Section 5.3.)

Given that no closures are projected to occur under the final rule and that EPA does not attempt to project production changes as a result of the rule, EPA estimates that no employment and other direct and indirect impacts will occur under this rule. Similarly, EPA concludes there will be no measurable local or national impacts in the commercial sector associated with closures. Should some facilities cut back operations as a result of this final regulation, EPA cannot project how great these impacts would be as it cannot identify the communities where impacts might occur (see Section 5.3).

Since EPA's closure analysis projects no facility or company closures under the final regulation, the Agency considers these final technology options to be economically achievable for commercial facilities (and companies).

5.1.1.2 Moderate Impacts

Some operations will likely incur additional moderate impacts, short of closure, see Table 5-1. EPA estimates that 4 commercial facilities incur costs greater than 5 percent of sales. This represents about 4 percent of all existing in-scope commercial facilities and approximately 6 percent of all existing in-scope facilities that are not projected to be baseline closures. No facilities have costs that exceed 10 percent of annual revenue.

EPA's analysis shows one company failing the credit test (which measures borrowing capacity) but no company experiencing a change in financial health as a result of the final regulation. This is based on EPA evaluation of the companies represented in the Agency's detailed questionnaire.

5.1.1.3 Sensitivity Analyses

As discussed in Section 3.2.1.5, EPA performed several sensitivity analyses based on the measure used for earnings (cash flow or net income), starting year for the projections (2001 or 2000), and any non-zero score considered a closure.[2] The results for the baseline closure analysis are:

- ■ (A) Earnings = Net income and starting year for projections = 2001
  Number of baseline closures is 43.

- ■ (B) Earnings = Cash flow, starting year for projections = 2001, and any score above zero is considered a closure. Number of baseline closures is 34.

- ■ (C) Earnings = Cash flow and starting year for projections = 2000
  Number of baseline closures is 27.

- ■ (D) Earnings = Net income and starting year for projections = 2000
  Number of baseline closures is 40.

These compare with the 32 baseline closures under standard methodology where earning are measured by cash flow and the starting year for projections is 2001.

EPA also examined the range of impacts under the final option with these sensitivity analysis. Under sensitivity analyses A and B, there are 2 incremental closures. Under sensitivity analyses C and D, there are no incremental closures.

Additionally, sensitivity analysis C allows EPA to assess impacts on an additional 10 facilities that were baseline closures in the primary analysis. These facilities reported at least one year of non-negative earnings. All 10 facilities are projected to remain open and none would incur any impacts as a result of the rule.

---

[2]The difference between cash flow and net income is that EPA adds in depreciation as a cost for the facility when calculating net income (i.e., the earnings for any given year will be lower under net income than they will be under cash flow, assuming the facility reports depreciation in the detailed survey). Cash flow is the primary basis for estimating closures as a result of the rule because it is a more accurate reflection of what is "in the facility's cash register" at the end of any given year, compared to what is reported for tax purposes. Net income is more conservative and a less objective measure (given the different ways a company or facility can report depreciation for tax purposes). Accounting references also recommend discounted cash flow. (See Appendix A)

The second parameter that is varied is the starting year for the earnings forecasts. As noted by several commenters, 2001 was a much less profitable year for the industry as a whole than was 2000.

The third parameter changes the closure decision. Closure is the most severe impact possible, and EPA therefore uses the "weight of evidence" approach to making that decision. Because there are three forecasting methods, the weight-of-evidence approach results in a facility being considered a closure when it shows negative long-term earnings under two or three of the forecasting methods (i.e., score =2). Changing the decision to a facility being considered a closure when it shows negative earnings under one or more forecasting methods (i.e., scores ≥1) dilutes the determination to identify situations where, if looked at in one particular manner, a site might show an impact.

5.1.2 Noncommercial Facilities

There are 141 noncommercial facilities within the scope of the rule. Of these, 141 are estimated to incur costs under the final regulation. The count represents Federal, State, Tribal (see Section 2.1 for a discussion of Tribal facilities), and Alaska nonprofit organizations. Based on the detailed questionnaire, EPA identified no academic/research facilities within the scope of the final rule.

In the absence of well-defined tests for projecting noncommercial facility closures, EPA compares pre-tax annualized compliance costs to 2001 operating budgets for noncommercial facilities. This analysis compares the incremental pollution control costs to the operating budget for the government facilities within the scope of the rule, and EPA conducts additional supplemental analysis of those surveyed facilities that report funding from user fees. A slightly different test is used for Alaska nonprofit facilities because they report revenues from harvested salmon. The comparison for Alaska nonprofit facilities is the pre-tax annualized cost to salmon revenues. More detailed discussion is provided in Section 3.3.

5.1.2.1 Budget Test

Objective measures for achievability are not available for public facilities. For Federal and State facilities, EPA compares the pre-tax annualized costs to the 2001 operating budget ("budget test"). EPA's analysis evaluates this test assuming a 3 percent, 5 percent, and 10-percent budget threshold.

Table 5-1 shows the effects on noncommercial operations from the final regulation based on EPA's economic analysis. Of the 141 noncommercial facilities, two are owned by Alaskan non-profits and are analyzed separately in Section 5.1.2.3 Of the remaining 139 noncommercial facilities, 4 facilities incur costs exceeding 10 percent of budget. EPA assumes that those facilities that face costs exceeding 10 percent of their budget would be adversely affected by the final regulation. These 4 facilities employ 16 people. None of these facilities report user fee funds; EPA could not conduct additional analyses to determine whether an increase in fees could offset these results. EPA's results, therefore, indicate that 3 percent of all non-commercial operations may be adversely affected by this final regulation. These operations may be vulnerable to closure based on the results of the Agency's budget test.

Under a 5-percent budget test, 12 facilities exceed the threshold under the final regulation. Among facilities that experience an increase in costs exceeding 5 percent, EPA assumes these facilities would face moderate financial impacts but would not be adversely affected. These results show that an additional 6 percent of all non-commercial operations (not counting those adversely affected) would experience some moderate impact associated with the costs of the rule. Some of these facilities report user fees revenues, see Table 5-1. Therefore, EPA conducts additional supplemental analyses to determine the magnitude of an increase in user fees could offset these results (see Section 5.1.2.2).

Given that the results of EPA's analysis projects that a small share of regulated noncommercial facilities may incur costs exceeding 10 percent of budget, estimated at 3 percent of facilities, the Agency considers these final technology options to be economically achievable for noncommercial facilities.

EPA maintains that there is potential for Tribal facilities to be present within the population of noncommercial facilities affected by this rule (see Section 2.1). EPA, recognizing that the mission of Tribal facilities may differ to some extent from the mission of State and Federally operated facilities, maintains that operating budgets, standardized for production level, are likely to be similar across all noncommercial facilities, including Tribal facilities. As such, the probabilities of adverse and moderate impacts among Tribal facilities are projected to be similar to those presented in Table 5-1 (approximately 3 percent and 9 percent, respectively). See Section 2.1 for discussion of Tribal facilities.

As part of analyses conducted prior to the Notice of Data Availability (NODA), EPA estimated impacts for Tribal facilities producing between 20,000 and 100,000 lbs/year for Option B (similar to the final Option) and identified no Tribal facility, represented in the detailed questionnaire, which incurred costs that exceeded 5 percent of budget (see also ERG, 2004). These results are for facilities that are not within scope of the final rule, but they provide additional evidence that the final rule is expected to be economically achievable for Tribal facilities.

5.1.2.2 User Fee Test

Table 5-2 provides the results of EPA's supplemental analysis that examines the extend to which government facilities, that fail a given budget test threshold, can recover increased costs through user fees. None of the facilities that fail a 10-percent budget test report funding from State user fees programs; therefore, EPA assumes that these facilities are not be able to raise user fees to offset compliance costs from this rule. However, 8 out of the 12 facilities that fail the 5% budget test reported that they use funds from user fees. These facilities would need to increase these funds by 7 percent to 9 percent to cover incremental compliance costs (see Table 5-2).

Section 3.3 presents information indicating that, when a state increases its fishing license fees (fees are not raised every year), increases typically range between 20 percent to 35 percent. In addition, on average, an increase in user fees by of 20 percent would raise fees to users by about $3 per user; an increase of 8 percent would be less than $1.50 per user (See Table 3-8 in Section 3.3 of this report.). These percent increases are not necessarily comparable to the percent increases in user fees needed to cover compliance costs, but this information still suggests that public facilities have opportunities to secure additional funding and/or alter management goals, that are not inconsistent with current management trends, to accommodate additional compliance costs such as those projected under this rule.

5.1.2.3 Alaska Nonprofit Facilities

EPA analyzed the impact of possible costs on Alaska nonprofit facilities by comparing the pre-tax annualized costs to reported salmon revenues. For the final rule, the costs were less than 0.2 percent of revenues.

Table 5-2
User Fee Analysis for Government Facilities

Budget Threshold	User Fee Increase	Number of Facilities
3%	Number of Facilities Failing Threshold	19
	Number of Facilities Not Reporting User Fee Funds	11
	<5 Percent[1]	0
	>5 Percent[1]	8
5%	Number of Facilities Failing Threshold	12
	Number of Facilities Not Reporting User Fee Funds	4
	<5 Percent[1]	0
	>5 Percent[1]	8
10%	Number of Facilities Failing Threshold	4
	Number of Facilities Not Reporting User Fee Funds	4
	<5 Percent[1]	0
	>5 Percent[1]	0

Numbers do not sum due to rounding.

[1] EPA's detailed survey of noncommercial facilities collected information on operating budgets and also requested that the respondent identify facility funding from fishing licenses, commercial fishing permits, vanity tags for vehicles, and special-purpose stamps. For the purpose of this analysis, EPA combined these funds under the general term "User Fees." The number of facilities that fail a test threshold and do not report user fee funds is reported on the line labeled "No User Fee." The other lines refer to whether a 5 percent increase in funding from user fees would or would not cover the estimated incremental pollution control costs.

5.1.3   Other Technology Options Considered by EPA

As described in Section 4.1, EPA considered a range of technology options during the development of this rulemaking, including Options A and B (discussed in NODA, USEPA, 2003) and Options 1, 2, and 3 (discussed at proposal USEPA, 2002a). This section presents the results of EPA's analysis across each of this regulatory options.

5.1.3.1   Commercial Facilities

Table 5-3 compares the results of the economic analysis of commercial facilities for the 5 regulatory options considered by EPA in addition to the final rule.[3] With regard to the closure analysis for commercial facilities, assuming cash flow and negative earnings for 2 out of 3 forecasts, EPA found no enterprise closures as a result of the rule under any option. For facilities, EPA identified no closures as a result of the rule under the final rule and Options A, B, 1, and 2. Four facilities are projected to close under Option 3. EPA identified no company closures as a result of the rule under the final rule and Options A, B, 1, and 2. One company is projected to close under Option 3.

With no projected facility closures under the final rule or Options A, B, 1, and 2, there are no associated losses in employment or increased local unemployment rates or national losses in employment or output (see Section 5.4). Under Option 3, the four facility closures result in a loss of 15 jobs. The lost jobs, in turn, result in increases of less than 1 percent in the local county unemployment rate. Under Option 3, the national employment loss is estimated to be 58 jobs. The estimated loss in output is $6.3 million in 2003 dollars.

EPA also estimated potential moderate impacts on commercial facilities under these options. Of the 69 facilities in the analysis (i.e., commercial facilities that are not baseline closures), 4 facilities fail a 5 percent threshold (final rule and Options B, 1, and 2). Under Option 3, an estimated 9 facilities fail a 5 percent sales test. No facilities fail a 10 percent threshold under any option considered. EPA projects no changes in financial health under the final rule and Options A, B, 1, and 2. EPA projects that one company is likely to change from favorable to vulnerable under Option 3. EPA projects no impacts on the borrowing capacity of the companies represented in the detailed questionnaire under Options A, B, 1, and 2. EPA projects that one company would have difficulty meeting the credit test under the final rule and Option 3.

EPA also performed a sensitivity analysis of varying O&M costs and for costs associated with activated carbon filtration. Documentation for these analyses is located in the rulemaking record OW-2002-0026 (ERG, 2003a and 2003b).

---

[3]The numbers in Table 5-3 differ from those presented in USEPA, 2003, Table VI.B.1 due to further refinements in the survey weights and clarifications on facility operations which became available after the NODA.

Table 5-3
Impacts for All Commercial Facilities, All Production Systems

Analysis Level	Number of Facilities or Companies for Which Analysis is Possible[1]	Impact	Option					
			Final	A	B	1	2	3
Enterprise	8	Closure	0	0	0	0	0	0
Facility	69	Closure	0	0	0	0	0	4
	69	Direct Employment Loss (lost jobs)	0	0	0	0	0	15
	69	Increase in County Unemployment (%)	0	0	0	0	0	<1 %
	69	National Employment Loss	0	0	0	0	0	50
	69	National Loss in Output ($ millions, 2003 dollars)	$0.0	$0.0	$0.0	$0.0	$0.0	$6.3
	69	Sales test >3%	4	0	4	4	4	9
	69	Sales test >5%	4	0	4	4	4	9
	69	Sales test >10%	0	0	0	0	0	0
Company	NA[2]	Closure	0	0	0	0	0	1
	NA	Farm Financial Health	0	0	0	0	0	A
	NA	Credit Test	1	0	0	0	0	1

[1]The number of facilities analyzed is equal to the number of in-scope facilities minus baseline closures. The number of companies analyzed is unweighted. Numbers of facilities in analysis and closures based on cash flow and negative earnings under 2 of 3 forecasts.
[2]Analysis performed at the company level. EPA evaluated 34 unweighted companies representing the 101 weighted facilities from the detailed questionnaire. The statistical weights, however, are developed on the basis of facility characteristics and therefore cannot be used for estimating the number of companies.
A: one company changes from favorable to vulnerable.

5.1.3.2 Noncommercial Facilities

Table 5-4 compares the results of the economic analysis of noncommercial facilities for the final rule and the 5 other regulatory options considered by EPA. Table 5-4 presents the findings for a 3, 5, and 10-percent budget threshold. The final rule shows impacts within the range represented by Option A and Option B and equal to or lower than the impacts for Options 1, 2, and 3. Option 3 shows the most facilities exceeding the 10 percent threshold.

Table 5-4 also shows the results of EPA's supplemental analysis of noncommercial facilities that report funding from user fees that are expected to incur costs exceeding EPA's budget test. Not all noncommercial facilities generate revenue from user fees. Under the 10 percent budget test, none of the affected facilities report user fee income under final rule and Options A, B, 1 and 2. The additional 4 facilities that fail the 10 percent budget test under Option 3 report user fee income and that income would need to increase more than 5 percent to cover the incremental pollution control costs.

Under the 5 percent budget test, the final rule is more flexible than Option A; 8 of 12 facilities have the potential to recover higher costs through increased user fees under the final rule while 0 of 7 facilities have that potential under Option A. As with the 10 percent budget test, the facilities would need to raise fees by more than 5 percent to compensate for the costs associated with the particular technology option.

5.1.4 Operations Producing Less than 100,000 lbs/yr

As part of the development of this final regulation, EPA also considered extending option requirements to existing operations that produce between 20,000 lbs/yr and 100,000 lbs/yr (see USEPA, 2002a and 2003). Section 5.1.4.1 provides a description of this group that are CAAP facilities but not within the scope of the rule. Section 5.1.4.2 provides a summary of EPA's regulatory analysis of the estimated impacts of Options A and B on facilities in this size category. More detailed information is the rulemaking record (ERG, 2004).

5.1.4.1 Description

There are approximately 257 facilities with production between 20,000 lbs/yr to 100,000 lbs/yr based on the detailed questionnaire compared to the estimated number of in-scope facilities is 242. Of these smaller facilities, 81 are commercial and 176 are noncommercial. Table 5-5 summarizes the number of commercial and noncommercial facilities by production system.

Of the 81 commercial facilities, 36 facilities (or 44 percent) report unpaid labor and or management. Thirty-five facilities (or 43 percent) are unprofitable in the facility closure analysis before the inclusion of incremental pollution control costs. The 81 (weighted) commercial facilities are represented by 16 unweighted companies[4] with the following organizational structure: 7 sole proprietorships, 2 partnerships (1 limited and 1 general), 4 S Corporations, and 3 C Corporations.

---

[4] A facility in the 20,000 to 100,000 lbs/yr category belongs to a 17th company that also owns in-scope (>100,000 lbs/yr) facilities. In order to avoid double-counting this company, it is not included in the count of companies in the 20,000 to 100,000 lbs/yr group.

Table 5-4
User Fee Analysis for Government Facilities

Budget Threshold	Estimated Number of Facilities	Number of Facilities					
		Final	Option A	Option B	Option 1	Option 2	Option 3
3%	Number Failing	19	11	26	19	23	45
	No User Fee	11	11	18	11	15	34
	<5 Percent	0	0	0	0	0	0
	>5 Percent	8	0	8	8	8	12
5%	Number Failing	12	7	15	15	15	24
	No User Fee	4	7	7	7	7	12
	<5 Percent	0	0	0	0	0	0
	>5 Percent	8	0	8	8	8	12
10%	Number Failing	4	3	7	7	7	11
	No User Fee	4	3	7	7	7	7
	<5 Percent	0	0	0	0	0	0
	>5 Percent	0	0	0	0	0	4

Numbers do not sum due to rounding.
[1] EPA's detailed survey of noncommercial facilities collected information on operating budgets and also requested that the respondent identify facility funding from fishing licenses, commercial fishing permits, vanity tags for vehicles, and special-purpose stamps. For the purpose of this analysis, EPA combined these funds under the general term "User Fees." The number of facilities that fail a test threshold and do not report user fee funds is reported on the line labeled "No User Fee." The other lines refer to whether a 5 percent increase in funding from user fees would or would not cover the estimated incremental pollution control costs.

Approximately 41 percent of the organizations are sole proprietorships. Of the 16 companies, all but one are small businesses. Overall, as compared with in-scope facilities, facilities that produce between 20,000 lbs/yr and 100,000 lbs/yr are more likely to be sole proprietorships, more dependent on unpaid labor and management, belong to a small business, and less likely to be profitable.

The 176 noncommercial facilities in the detailed questionnaire data that produce between 20,000 lbs/yr to 100,000 lbs/yr include: 154 Government facilities; 7 Alaska nonprofit organizations; one Academic/research facility; and 14 Tribal facilities. That is, operations with between 20,000 lbs/yr to

100,000 lbs/yr group of facilities encompasses an additional type of organizations—academic/research facilities.

Table 5-5
Number and Types of Facilities with Annual Production between 20,000 and 100,000 Pounds

Production System	Owner	Estimated Number of Facilities				
		Initial Facility Count	Baseline Closures	In Analysis [1]	In Analysis that Incur Costs	In Cost Totals [2]
Flow Through and Recirculating	Commercial	81	35	46	41	47
	Non-commercial	175	NA	175	175	175
Net Pen	Commercial	0	0	0	0	0
	Non-commercial	1	NA	1	0	1
Total	Commercial	81	35	46	41	47
	Non-commercial	176	NA	176	176	176

Totals may not sum due to rounding
NA: not applicable.
[1] In analysis counts are calculated by subtracting out baseline closures from the initial facility count.
[2] Start-up operations are in the cost totals but have insufficient information to be in the economic analysis.

5.1.4.2 Economic Impact Analysis

For comparison purposes EPA conducted an analysis of the regulatory impacts under two regulatory options (Option A and Option B) for this size group (production between 20,000 lbs/yr to 100,000 lbs/yr).[5]

For commercial facilities, EPA's analysis indicates that all facilities in the 20,000 lb/yr to 100,000 lb/yr category that are financially healthy enough to pass the baseline analysis are healthy enough to remain open under either option. None of the commercial companies experience a change in financial health or suffer impaired credit. No operations fail the 5 percent sales test threshold. For noncommercial facilities, EPA's analysis indicates that a substantially higher number of operations would fail a 10-percent budget test, estimated at 8 facilities (Option A) and 23 facilities (Option B) or 5 percent and 13 percent, respectively. This compares to 3 percent and 9 percent of facilities that may be adversely impacted among operations that produce more than 100,000 lbs/yr. No Tribal, academic/research, and Alaska nonprofit facility fails a 5-percent budget test. More detailed information is the rulemaking record (ERG, 2004).

---

[5]These facilities are not within the scope of the final rule and, thus, final rule costs were not estimated for them.

5.2   NEW SOURCE PERFORMANCE STANDARDS (NSPS)

To evaluate potential effects to new aquaculture facilities, EPA examines possible barriers to entry to a new facility because of the final regulation. First, EPA examines the proportion of commercial facilities that incur no costs under each option. About 4 percent of regulated facilities do not incur any costs under the final regulation. Second, EPA examines the proportion of commercial facilities with no land or capital costs under each option. About 76 percent of facilities incur no land or capital costs. Third, for the subset of companies with incremental land or capital costs, EPA examines the ratio of those costs to total company assets. This comparison is calculated on company data because asset data were collected only at the company level. (Facility weights cannot be used for the company analyses.) EPA calculates the ratio for each company and took the average of the ratios. The incremental land and capital costs, where they were incurred, represented less than 0.2 percent of total assets. Based on these results, the final regulation does not appear to present a barrier to entry to new operations.

EPA also evaluated the regulatory analysis of new source facilities with production between 20,000 lbs/yr to 100,000 lbs/yr. For comparison purposes EPA conducted an analysis of the regulatory impacts under two regulatory options (Option A and Option B). About 10 percent of expected new facilities in this size category incur no costs under the final regulation. EPA's analysis examines the proportion of commercial facilities with no land or capital costs under each option. Nearly two-thirds of the facilities incur no land or capital costs under the final regulation. Among facilities that incur costs, these costs are annual costs rather than land or capital for two of every three facilities. The incremental land and capital costs, where incurred, account for about 1.5 percent of total assets under the final regulation. More detailed information is the rulemaking record (ERG, 2004).

5.3   MARKET AND FOREIGN TRADE IMPACTS

5.3.1   Market Impacts

EPA was not able to prepare a market model analysis for this rule for reasons described in Section 3.6 of this report. Because EPA was not able to prepare a market model analysis for this rule, the Agency is not able to report quantitative estimates of changes in overall supply and demand for aquaculture products and changes in market prices, as well as changes in traded volumes including imports and exports. EPA examined the impacts two ways. In the first or base analysis, no costs are passed through to the consumer and all impacts fall on commercial facilities (i.e., conservative approach for closure analysis). In the second case, all costs are assumed to be passed to the consumer and no impacts fall on the commercial facilities. As a result of comparing the results of the analyses, EPA does not expect significant market impacts as a result of this final regulation.

For closure analysis, results show that no commercial facilities are projected to close under a "no cost pass-through" assumption. About 3 percent of all noncommercial facilities might experience adverse financial effects associated with the rule (Section 5.1). These estimated impacts coupled with the overall cost of the rule, as compared to the total value of the U.S. aquaculture industry, lead EPA to believe that the effects of this regulation on U.S. aquaculture markets will be modest.

To approximate potential maximum market impacts on the consumer under this final rule, EPA performed a bounding analysis on prices if all costs were passed through to the consumer. Under this scenario, there would be no impacts on commercial facilities because all costs were passed through to the consumer. The estimated pre-tax cost of the final rule to in-scope commercial facilities that would be passed to the customer is $0.279 million in 2003 dollars (based on estimates shown in Table 4-3). The amount of 2001 production from in-scope flow-through, netpen, and recirculating commercial facilities is 94 million pounds (see Table 2-6). If all costs are assumed to be passed through, the typical price per pound would increase, at most, by 0.3 cents per pound because of this final regulation.

5.3.2 Foreign Trade Impacts

Although foreign trade impacts are difficult to predict, since agricultural exports are determined by economic conditions in foreign markets and changes in the international exchange rate for the U.S. dollar, EPA does not expect significant changes in net trade as a result of this final regulation. EPA projects no rule-induced closures as a result of this rule. EPA also believes that long-term shifts in supply associated with this rule are unlikely given competition from domestic wild harvesters and foreign suppliers.

As discussed in Section 3.6 of this report, the U.S. is not a major player in world aquaculture markets, accounting for about 1 percent of world production by weight. Due to the relatively small market share of U.S. aquaculture producers in world markets, EPA believes that long-term shifts in supply associated with this rule are unlikely given expected continued competition from domestic wild harvesters and foreign suppliers. Although increased costs of this final regulation could exacerbate competitive pressures that are already facing U.S. aquaculture producers, EPA believes that any future widening of the current trade gap between U.S. imports and exports will be mostly attributable to existing market influences beyond the cost of this final regulation. This is based on information on the current competitive role of the U.S. in world aquaculture markets and also expectations that consumer aquaculture demand in the U.S. will continue to outpace U.S. domestic production. This is confirmed by an FAO study of projected changes in U.S. aquaculture production and net trade from 1997 to 2030 indicating modest increases in U.S. production but an increase in net imports, mostly attributable to rising consumer demand. EPA concludes therefore that the impact of this final rule on U.S. aquaculture trade will not be significant.

Current competitive pressures facing U.S. aquaculture producers might also be challenged through other U.S. governmental programs that are designed to address concerns about competition to U.S. farmers from lower-cost world producers. For example, the 2002 Trade Act (Public Law 107-210) established the Trade Adjustment Assistance for Farmers program. Under this program—administered by USDA's Foreign Agricultural Service (FAS)—U.S. agricultural producers (including those that raise aquatic animals) may be eligible for technical assistance and a financial payment, if they believe they have suffered from low prices due to increasing imports.

5.4 REFERENCES

ERG (Eastern Research Group). 2004. Technical Directive No. 3, Items 1 and 2. Description and Economic Impact Analysis of CAAP Facilities that Produce Between 20,000 to 100,000 lb/yr in Flow-Through, Recirculating, and Net Pen Systems. Memorandum to Chris Miller and Renee Johnson, EPA. May 6.

ERG (Eastern Research Group). 2003a. "BMP Option: Preliminary Results from 23 June 2003 Aquatic Production Costs." Memorandum to Chris Miller, EPA. 27 June. Docket OW-2002-0026 DCN 20410.

ERG (Eastern Research Group). 2003b. "CAAP Sensitivity Analysis: Activated Carbon Cost." Memorandum to Chris Miller, EPA. 10 October. Docket OW-2002-0026 DCN 20443.

USEPA (U.S. Environmental Protection Agency). 2003. Effluent Limitations Guidelines and New Source Performance Standards for the Concentrated Aquatic Animal Production Point Source Category; Notice of Data Availability; Proposed Rule. 40 CFR Part 451. *Federal Register* 68:75068-75105. December 29.

USEPA (U.S. Environmental Protection Agency). 2002a. Effluent Limitations Guidelines and New Source Performance Standards for the Concentrated Aquatic Animal Production Point Source Category; Proposed Rule. 40 CFR Part 451. *Federal Register* 67:57872-57928. September 12.

USEPA (U.S. Environmental Protection Agency). 2002b. Economic and Environmental Impact Analysis of the Proposed Effluent Limitations Guidelines and Standards for the Concentrated Aquatic Animal Production Industry. Washington, DC. EPA-821-R-002-015. September.

CHAPTER 6

EVALUATION OF THE EFFECTS OF THE RULE ON SMALL ENTITIES

This section considers the effects of the regulations on small businesses. Section 6.1 discusses EPA's requirements under the Regulatory Flexibility Act. Section 6.2 outlines EPA's initial assessment of small businesses in the sectors affected by the regulations. Section 6.3 describes the EPA's compliance with RFA requirements and Section 6.4 presents the analysis of economic impacts to small entities that are affected by the final regulation.

6.1 THE REGULATORY FLEXIBILITY ACT AS AMENDED BY THE SMALL BUSINESS REGULATORY ENFORCEMENT FAIRNESS ACT

The Regulatory Flexibility Act (RFA, 5 U.S.C et seq., Public Law 96-354) as amended by the Small Business Regulatory Enforcement Fairness Act of 1996 (SBREFA) generally requires an agency to prepare a regulatory flexibility analysis describing the impact of the regulatory action on small entities as part of the rulemaking. This is required of any rule subject to notice and comment rulemaking requirements under the Administrative Procedure Act or any other statute unless the agency certifies that the rule will not have a "significant impact on a substantial number of small entities." Small entities include small businesses, small organizations, and governmental jurisdictions. The RFA acknowledges that small entities have limited resources and makes it the responsibility of the regulating Federal agency to avoid burdening such entities unnecessarily. If, based on an initial assessment, a regulation is likely to have a significant economic impact on a substantial number of small entities, the RFA requires a regulatory flexibility analysis.

In addition to the preparation of an analysis, the RFA, as amended by SBREFA, imposes certain responsibilities on EPA when the Agency proposes rules that might have a significant impact on a substantial number of small entities. These include requirements to consult with representatives of small entities about the proposed rule. The statute requires that, where EPA has prepared an initial regulatory flexibility analysis (IRFA), the Agency must convene a Small Business Advocacy Review (SBAR) Panel for the proposed rule to seek the advice and recommendations of small entities concerning the rule. The panel is composed of employees from EPA, the Office of Information and Regulatory Affairs within the Office of Management and Budget, and the Office of Advocacy of the Small Business Administration (SBA).

EPA is certifying that this final regulation will not have a significant economic impact on a substantial number of small entities. Despite this determination, EPA has prepared an evaluation of the effects on small entities that examines the impact of the rule on small entities along with regulatory alternatives that could reduce that impact. EPA also prepared an economic analysis of the potential impacts to affected small businesses. For the 2002 Proposal, EPA prepared an IRFA, which was published in the *Federal Register* (USEPA, 2002a; see FR 67: 57916-57917) and presented as part of the Economic and Environmental Impact Analysis (EEIA) for the proposed rule (USEPA, 2002b).

6.2   INITIAL ASSESSMENT

Prior to the 2002 Proposal, EPA conducted an initial assessment according to Agency guidance on implementing RFA requirements (USEPA, 1999). First, EPA must indicate whether the proposal is a rule subject to notice-and-comment rulemaking requirements. EPA determined that the proposed regulation is subject to notice-and-comment rulemaking requirements. Second, EPA should develop a profile of the affected small entities. EPA has developed such a profile of the aquaculture industry, which includes all affected operations as well as small businesses. This industry profile is provided in the Proposal EEIA (USEPA, 2002b, Chapter 2). Third, EPA's assessment needs to determine whether the rule would affect small entities and whether the rule would have an adverse economic impact on small entities.

For the proposed rulemaking, EPA concluded that costs are sufficiently low to justify "certification" that the regulations would not impose a significant economic impact on a substantial number of entities (USEPA, 2002a; see FR 67: 57916). In addition, however, EPA also complied with all RFA provisions and conducted outreach to small businesses, convened an SBAR Panel, and prepared an IRFA. That analysis described EPA's assessment of the impacts of the proposed regulations on small businesses in the aquaculture industry. A summary of this analysis was published in the *Federal Register* at the time of publication of the 2002 Proposal (USEPA, 2002a; see FR 67: 57916-57917). More detailed information on EPA's IRFA is provided in the Proposal EEIA (USEPA, 2002b, Section 8.3). EPA's Proposal EEIA also describes other requirements of EPA's initial assessment of small businesses and summarizes the steps taken by EPA to comply with the RFA (USEPA, 2002b, Section 8.4).

6.2.1   Definitions of a Small Aquaculture Entity

The RFA/SBREFA defines several types of small entities, including small governments, small organizations, and small businesses.

A "small governmental jurisdiction" is defined as the government of a city, county, town, school district, or special district with a population of less than 50,000. For the purposes of the RFA, Federal, State, and Tribal governments are not considered small governmental jurisdictions (USEPA, 1999).[1] Federal facilities, regardless of their production levels, are not part of small governments. EPA identified no public aquaculture facilities belonging to small governments that are affect by the rule. EPA identified our small organization, an Alaska nonprofit, within the scope of the rule.

The Small Business Administration (SBA) sets size standards to define whether a business entity is small and publishes these standards in 13 CFR 121. The standards are based either on the number of employees or annual receipts. Table 6-1 lists the North America Industry Classification System (NAICS) codes potentially in scope of the proposed rule and their associated SBA size standards as of January 1, 2002 (SBA, 2000 and SBA, 2001).

---

[1] See Section 9 of this report where impacts on these entities are summarized in accordance with Unfunded Mandates Reform Act (UMRA) requirements.

Table 6-1
Small Business Size Standards

NAICS Code	Description	Size Standard (Annual Revenues)
112511 112519	Finfish Farming and Fish Hatcheries Other Animal Aquaculture	$0.75 million $0.75 million

When making classification determinations, SBA counts receipts or employees of the entity and all of its domestic and foreign affiliates (13 CFR.121.103(a)(4)). SBA considers affiliations to include: stock ownership or control of 50 percent or more of the voting stock or a block of stock that affords control because it is large compared to other outstanding blocks of stock (13 CFR 121.103(c)); common management (13 CFR 121.103(e)); and joint ventures (13 CFR 121.103(f)).

EPA assumes the following for its evaluation:

- Sites with foreign ownership are not small (regardless of the number of employees or receipts at the domestic site).

- The definition of small is set at the highest level in the corporate hierarchy and includes all employees or receipts from all members of that hierarchy.

- If any one of a joint venture's affiliates is large, the venture cannot be classified as small.

6.2.2 Number of Small Businesses Affected by the Final Regulation

Based on detailed questionnaire data, EPA identified 37 facilities belonging to small businesses. It is quite possible for a small facility to belong to a large business, but a large facility—by definition—must belong to a large business.

6.2.3 Results of the Initial Assessment for the 2002 Proposal

For past regulations, EPA has often analyzed the potential impacts to small businesses by evaluating the results of a costs-to-sales test, measuring the number of operations that will incur compliance costs at varying threshold levels (including ratios where costs are less than 1 percent, between 1 and 3 percent, and greater than 3 percent of gross income). EPA conducted such an analysis at the time of the 2002 proposal, indicating that roughly 30 percent of the estimated number of small businesses directly subject to the rule might incur costs in excess of three percent of sales.

EPA's initial assessment at proposal covers facilities that produce more than 100,000 lbs/yr and met SBA's small business definition, consisting of 36 commercial facilities and 12 Alaska facilities (belonging to 8 nonprofit organizations). The results of this initial assessment indicate that 17 of 36 commercial facilities failed the 1 percent sales test (cost-to-sales ratio) and 10 of 36 commercial facilities failed the 3 percent sales test. The maximum cost-to-sales ratio among these facilities was 7 percent. Among the Alaska nonprofit organizations, 3 of 6 facilities failed a 1 sales test, and 1 of 6 facilities failed the 3 percent sales test. A summary of this analysis was published as part of the proposed rule (USEPA, 2002a; see FR 67: 57916-57917), with more detailed information provided in the Proposal EEIA (USEPA, 2002b, Section 8.3).

6.3   RFA _____ AND SBAR PANEL

6.3.1   Outreach and Small Business Advocacy Review

EPA's engaged in outreach activities and convened a SBAR Panel to obtain the advice and recommendations of representatives of the small entities that potentially would be subject to the rule's requirements. The Agency convened the SBAR Panel on January 22, 2002. Members of the Panel represented the Office of Management and Budget, the Small Business Administration and EPA. The Panel met with small entity representatives (SERs) to discuss the potential effluent guidelines and, in addition to the oral comments from SERs, the Panel solicited written input. In the months preceding the Panel process, EPA conducted outreach with small entities that would potentially be affected by the Agency's CAAP regulation. On January 25, 2002, the SBAR Panel sent some initial information for the SERs to review and provide comment. On February 6, 2002 the SBAR Panel distributed additional information to the SERs for their review. On February 12 and 13, the Panel met with SERs to hear their comments on the information distributed in these mailings. The Panel also received written comments from the SERs in response to the discussions at this meeting and the outreach materials. The Panel asked SERs to evaluate how they would be affected and to provide advice and recommendations regarding early ideas to provide flexibility. See Section 8 of the Panel Report for a complete discussion of SER comments. The Panel evaluated the assembled materials and small-entity comments on issues related to the elements of the IRFA. A copy of the Panel report is included in the docket for this proposed rule (see DCN 31019). EPA provided responses to the Panel's most significant findings as part of the proposed rule (USEPA, 2002a, 67: 57918-57920).

6.4   EVALUATION OF EFFECTS ON SMALL ENTITIES

EPA is certifying that this final regulation will not have a significant economic impact on a substantial number of small entities. EPA has evaluated the effects of the final rule on small entities, however, this review examines the impact of the rule on small entities along with regulatory alternatives that could reduce that impact. EPA's conclusions about potential impacts to affected small business of this rule are presented in Section 6.5.

6.4.1   Need for and Objectives of the Final Regulation

EPA is considering this action because aquaculture facilities may introduce a variety of pollutants into receiving waters. Under some conditions, these pollutants can be harmful to the environment and have a negative impact on water quality (Fries and Bowles, 2002; Loch et al., 1996; and Virginia, 2002). According to USDA's *1998 Census of Aquaculture*, there are approximately 4,000 commercial aquatic animal production facilities in the United States (USDA, 2000). Aquaculture has been among the fastest-growing sectors of agriculture until a recent slowdown that began several years ago. EPA analysis indicates that many aquaculture facilities have treatment technologies in place that greatly reduce pollutant loads. However, in the absence of treatment, pollutant loads from individual facilities, such as those covered by the rule, can contribute substantial amounts of nitrogen, phosphorus, and TSS per year to the receiving water body. These pollutants can contribute to eutrophication and other aquatic ecosystem responses to excess nutrient loads and BOD effects.

Another area of potential concern relates to non-native species introductions from aquaculture facilities, which may pose risks to native fishery resources and wild native aquatic species from the establishment of escaped individuals (Hallerman and Kapuscinski, 1992; Carlton, 2001; Volpe et al., 2000; Leung et al., 2002; and Kolar and Lodge, 2002). Aquaculture facilities also employ a range of drugs and chemicals used therapeutically that may be released into receiving waters. For some investigational drugs, as well as for certain application of approved drugs, there is a concern that further information is needed to fully evaluate risks to ecosystems and human health associated with their use in some situations (USEPA, 2002a). Finally, aquaculture facilities also may inadvertently introduce pathogens into receiving waters, with potential impacts on native biota. This final regulation addresses a number of these concerns. These regulations are proposed under the authority of Section 301, 304, 306, 308, 402, and 501 of the Clean Water Act, 33 U.S.C.1311, 1314, 1316, 1318, 1342, and 1361.

6.4.2 Significant Comments in Response to the IRFA

EPA responded to significant comments on the proposed rule and its initial regulatory flexibility analysis in the Notice of Data Availability (USEPA, 2003). The majority of these comments express concern over the ability of regulated facilities to absorb additional operating costs due to regulation, given that USDA's *1998 Census of Aquaculture* reports that over 96 percent of trout farms are small businesses. USDA's comparatively high estimate of the number of small farm businesses is due to differences between USDA's and SBA's definition. For example, SBA's size standards differ from the revenue cutoff generally recognized by USDA, which has set $250,000 in gross sales as its cutoff between small and large family farms (USDA, 1998).

EPA responded by using the detailed questionnaire data to capture revenue information at the facility and company level in order to identify small businesses; however, EPA continues to use SBA's small business definition per its guidance on how to comply with RFA/SBREFA requirements. EPA also presents a more thorough discussion of some of the other issues raised in public comments by conducting additional sensitivity analyses (e.g., cash flow, depreciation, sunk costs, capital replacement, and unpaid labor and management). See Appendix A for a more complete discussion of these topics.

6.4.3 Description and Estimate of Number of Small Entities Affected

Based on the information collected in its detailed questionnaire. Of the 38 facilities identified by EPA one is a noncommercial hatchery belonging to an Alaskan non-profit and 37 are commercial facilities belonging to small businesses. Of these, 36 are facilities in the Flow Through and Recirculating Subcategory and 2 are in the Net Pen Subcategory.

For the proposed rule, EPA stated its intention to make its final determination of the impact of the rule on small businesses based on analysis of detailed questionnaire data. However, EPA also convened a Small Business Advocacy Review Panel pursuant to RFA/SBREFA Section 609(b) (USEPA, 2002a, p.

2002a, p. 57909). For its 2003 Notice of Data Availability, EPA identified 117 facilities belonging to small businesses, seven facilities belonging to small organizations, and one academic/research facility among the facilities that produced more than 20,000 lbs/yr (USEPA, 2003). By restricting the scope of the rule to facilities that produce more than 100,000 lbs/yr, EPA also limited the number of small entities within the scope of the rule to 37 commercial facilities and one organization. The small business economic screening analysis for these 38 facilities is presented in Section 6.4.

6.4.4 Description of the Reporting, Recordkeeping, and Other Requirements

EPA's final rule includes a requirement for reporting Investigational New Animal Drugs (INADs) and extra-label use drugs, and a requirement to report failures and material damage to the structure of the aquatic animal containment system leading to a material discharge or pollutants . In addition to the BMP plan, the final regulation requires record keeping in conjunction with implementation of a feed management system. Flow through and recirculating facilities subject to the rule must record the dates and brief descriptions of rearing unit cleaning, inspections, maintenance and repair. Net pen facilities must keep the same types of feeding records as described above and record the dates and brief descriptions of net changes, inspections, maintenance and repairs to the net pens.

EPA estimates that each plan will require 40 hours per facility to develop the plan. The plan will be effective for the term of the permit (5 years). EPA assumed that each employee at a facility would incur a one time cost of 4 hours for initial BMP plan review. EPA included an annual cost for four hours of management labor to maintain the plan and eight hours of management labor and 4 hours for each employee for training and an annual review of BMP performance. EPA does not believe that the development and implementation of these BMPs will require any special skills. All of the CAAP facilities within the scope should currently be permitted, so incremental administrative costs of the regulation are negligible. However, Federal and State permitting authorities will incur a burden for tasks such as reviewing and certifying the BMP plan and reports on the use of drugs and chemicals. EPA estimated these costs at approximately $13,176 for the three-year period covered by the information collection request or roughly $4,392 per year.

6.4.5 Steps Taken to Minimize Significant Economic Impacts on Small Entities

EPA took several steps to minimize the potential impact of this final regulation. EPA restricted the rule in three major ways. First, EPA is restricting the rule to CAAP facilities rather than all facilities that raise aquatic animals. Second, EPA is restricting the scope of the rule to flow-through, recirculating, and net pen production systems. Third, EPA is restricting the scope to facilities that produce more than 100,000 lbs/yr. The USDA Census of Aquaculture identified approximately 4,000 aquaculture facilities nationwide (USDA, 2000). The final rule applies to an estimated 101 commercial facilities, approximately 2.5 percent of the total population.

Finally, EPA based the final rule on a technology option that has no adverse economic impacts on commercial facilities. While some commercial facilities may experience moderate impacts, EPA projects that no small businesses will close as a result of today's final rule. Given the results of this economic analysis of the effects on small businesses, EPA is certifying that this action will not have a significant economic impact on a substantial number of small entities.

6.4.6 Identification of Relevant Federal Rules that May Duplicate, Overlap, or Conflict with the Final Rule

Since the start of the rulemaking effort, Congress and Federal agencies have been working to clarify roles regarding the final CAAP regulation. EPA met with various stakeholders to ensure that other Federal rules would not duplicate, overlap, or conflict with the final rule.

EPA met with USDA's Animal and Plant Health Inspection Service (APHIS) to discuss how the requirements and objective of the CAAP rule relate to authorities under their jurisdiction (DCN 31123). At that meeting, USDA discussed how the Animal Health Protection Act (enacted as part of the 2002 Farm Bill), which gives APHIS the authority to develop and implement aquatic animal health programs. This law gives authority to APHIS for aquatic farm-raised animal disease management including emergency responses actions to invasive pathogen outbreaks. APHIS is also authorized to implement control programs using drugs or chemicals and biosecurity practices to reduce disease risk and impact on the industry. EPA and APHIS also discussed APHIS' broad mandate to address import and interstate movement of exotic species under the Federal Plant Pest Act and the Plant Quarantine Act.

EPA met with FDA to clarify FDA's environmental assessment requirements for the substances over which FDA has jurisdiction (DCN 31126). EPA and FDA are working on a formal agreement that would address environmental concerns about the discharge of drugs used at aquatic animal production facilities. This agreement, which might help protect the aquatic environment from harm, would facilitate information sharing about effluent concentrations of active drug ingredients. When appropriate, FDA would include in the labeling of approved new animal drugs, effluent concentrations of the active drug ingredient which should not be exceeded in wastewater discharges. EPA would notify permitting authorities who would incorporate these effluent concentrations into the NPDES permits as enforceable requirements.

6.5 EPA'S EVALUATION OF SMALL ENTITY IMPACTS

EPA's evaluation shows that the final rule will have no adverse economic impacts on commercial facilities, including small businesses. The results of EPA's economic analysis presented in Section 5.2 covers all regulated facilities, including both small business and businesses that do not meet SBA's small business definition. EPA projects no closures for facilities owned by small businesses. For the small organizations costs are less than 0.2 percent of salmon revenues.

EPA projects that this rule will have some moderate impacts on small businesses. One small business is projected to fail the credit test although no small companies undergo a change in financial status under the Financial Health Test described in Section 3.2.2.3 as a result of the rule. Four facilities belonging to small businesses have costs-to-sales ratios in excess of five percent. All four of these facilities are in the Flow Through and Recirculating Subcategory. No facilities have costs between 3 percent and 5 percent of sales.

6.6   REFERENCES

Carlton, J.T. 2001. *Introduced Species in U.S. Coastal Waters. Environmental Impacts and Management Priorities.* Prepared for the Pew Oceans Commission, Arlington, VA., 28 pp.

Fries, L.T. and D.E. Bowles.. 2002 Water Quality and Macroinvertebrate Community Structure Associated with a Sportfish Hatchery Outfall. *North American Journal of Aquaculture* 64: 257-266. EPA Docket No. OW-2002-026, DCN 40621.

Hallerman, E.M., and A.R. Kapuscinski, 1992. Ecological Implications of Using Transgenic Fishes in Aquaculture. *ICES March Science Symposium* 194:56-66.

Kolar, C.S. and D.M. Lodge. 2002. Ecological Predictions and Risk Assessment for Alien Fishes in North America. *Science* 298:1233-1236. EPA Docket No. OW-2002-026, DCN 40569.

Leung, B., D.M. Lodge, D. Finoff, J.F. Shogren, M.A. Lewis, and G. Lamberti. 2002. An Ounce of Prevention or a Pound of Cure: Bioeconomic Risk Analysis of Invasive Species. *Proceedings of the Royal Society of London, Series B* 269: 2407-2413. EPA Docket No. OW-2002-026, DCN 40568.

Loch, D.D., J.L. West, and D.G. Perlmutter. 1996. The Effect of Trout Farm Effluent on the Taxa Richness of Benthic Macroinvertebrates. *Aquaculture* 147: 37-55. EPA Docket No. OW-2002-026, DCN 61497.

SBA (Small Business Administration). 2001. 13 CFR Parts 107 and 121 Size eligibility requirements for SBA financial assistance and size standards for agriculture. Direct Final Rule. 65 FR 100:30646-30649. 7 June.

SBA (Small Business Administration). 2000. 13 CFR Part 121 Small business size regulations: Size standards and the North American Industry Classification System; Final Rule. 65 FR 94:30836-30863. 15 May.

USDA (United States Department of Agriculture). 2000. *1998 Census of Aquaculture.* Also cited as 1997 Census of Agriculture. National Agricultural Statistics Service. Volume 3, Special Studies, Part 3. AC97-SP-3. February.

USDA (U.S. Department of Agriculture). 1998. Report of the USDA National Commission on Small Farms: A Time to Act. MP-1545. January. http://www.reeusda.gov/agsys/smallfarm/report.htm

USEPA (U.S. Environmental Protection Agency). 2003. Effluent Limitations Guidelines and New Source Performance Standards for the Concentrated Aquatic Animal Production Point Source Category; Notice of Data Availability; Proposed Rule. 40 CFR Part 451. *Federal Register* 68:75068-75105. December 29.

USEPA (U.S. Environmental Protection Agency). 2002a. Effluent Limitations Guidelines and New Source Performance Standards for the Concentrated Aquatic Animal Production Point Source Category; Proposed Rule. 40 CFR Part 451. *Federal Register* 67:57872-57928. September 12.

USEPA (U.S. Environmental Protection Agency). 2002b. Economic and Environmental Impact Analysis of the Proposed Effluent Limitations Guidelines and Standards for the Concentrated Aquatic Animal Production Industry. Washington, DC. EPA-821-R-002-015. September.

USEPA (U.S. Environmental Protection Agency). 1999. *Revised Interim Guidance for EPA Rulewriters: Regulatory Flexibility Act as amended by the Small Business Regulatory Enforcement Fairness Act.* Washington, DC. 29 March. EPA Docket No. OW-2002-026, DCN 20121.

Virginia (The Virginia Water Resources Research Center). 2002. *Benthic TMDL Reports for Six Impaired Stream Segments in the Potomac-Shenandoah and James River Basins.* Prepared for the Virginia Department of Environmental Quality, VA Department of Conservation and Recreation. EPA Docket No. OW-2002-026, DCN 40571. http://www.deq.state.va

Volpe, J.P., E.B. Taylor, D.W. Rimmer, and B.W. Glickman. 2000. Evidence of Natural Reproduction of Aquaculture-Escaped Atlantic Salmon in a Coastal British Columbia River. *Conservation Biology.* 14(June):899-903.

CHAPTER 7

ENVIRONMENTAL IMPACTS FROM AQUACULTURE FACILITIES

7.1   INTRODUCTION

The purpose of this chapter is to present information EPA has collected relating to environmental impacts from aquaculture facilities, with a focus on the larger concentrated aquatic animal production (CAAP) facilities that are in the scope of EPA's final CAAP rule. Environmental effects associated with types of production systems and segments of the industry that are not in the scope of EPA's final rule (e.g., pond systems; molluscan shellfish operations) are addressed to a very limited extent by this chapter. In addition, EPA has not attempted to prioritize or otherwise characterize environmental risks from any particular impact, nor has EPA attempted to review in this chapter industry, State, and other regulations and programs to mitigate potential environmental impacts from CAAP facilities (see Chapter 1 of the Technical Development Document for a discussion of existing regulations affecting this industry).

A summary overview of CAAP pollutant loadings, including a brief review of facility characteristics, effluent quality, and range of annual pollutant loadings, is presented in Section 7.2. A limited review of selected literature relating to the water quality and aquatic ecosystem impacts from these loadings is presented in Section 7.3. References are provided in Section 7.4. The sources cited in this chapter include EPA engineering analyses that can be found in the Technical Development Document (USEPA, 2004) accompanying EPA's final CAAP rule; materials submitted with public comments and other materials provided by stakeholders; and a range of published technical literature.

7.2   CAAP INDUSTRY DISCHARGES

7.2.1   Description of Industry

The aquaculture industry encompasses several major types of production systems and a wide range of sizes and species. According to EPA census data, there are over 3,200 aquatic animal production systems in the United States, with approximately 260 facilities subject to EPA's final regulation. Effluent quality varies with facility characteristics including type of production system, facility size, and ownership.

Aquatic animal production facilities that are in-scope of the final CAAP rule represent considerable variation in facility size, species, production system type, ownership, and geographic distribution. The size of in-scope facilities varies by annual production levels. Aquatic animal production facilities produce a variety of species in a number of different production systems, including ponds, flow-through systems, recirculating systems, net pens, and open water culture. Furthermore, aquatic animal production facilities are owned by commercial and non-commercial (e.g., state and federal governments, tribes, non-profits, and research institutions) entities and vary in their location throughout the United States. Refer to Chapter 3 of the Technical Development Document for a detailed summary of the in-scope facilities.

7.2.2 Discharges of Solids, Nutrients, and BOD

7.2.2.1 Introduction

Solids, nutrients, and BOD primarily arise from uneaten feed and waste produced by the fish. A number of earlier investigations to characterize aquaculture effluents have been performed (e.g., as described in Regional Aquaculture Centers, (RAC), 1998. The following sections focus on characterizing concentrations from studies reported in the literature and EPA sampling observations. The following examples are representative of in-scope facilities because they examine facilities that are similar to facilities in-scope of the final CAAP rule, in terms of size, production systems, and general operation. A later section provides estimates of annual mass loadings from facilities that are in-scope of the final CAAP rule.

7.2.2.2 Flow-through Systems

Effluents from flow-through systems can be characterized as continuous, high-volume flows containing low pollutant concentrations. Effluents from flow-through systems are affected by whether a facility is in normal operation or whether the tanks or raceways are being cleaned. Waste levels can be considerably higher during cleaning events (Hinshaw and Fornshell, 2002; Kendra, 1991).

Hinshaw and Fornshell (2002) compiled effluent values reported in the literature and provide ranges for various water quality constituents. They report average BOD levels to be 2.0 mg/L during normal operations, with levels increased by approximately 10 times as settleable solids were disturbed during cleaning. Likewise, solids increased from normal levels of ≤35 mg/L to a range of 61.9-1000 mg/L for facilities during cleaning. Concentrations of total phosphorus (TP) reported were ≤0.13 mg/L, but increased by three times during cleaning[2]. Estimates of ammonia-nitrogen ranged from 0.01 to 1.52 mg/L, illustrative of the fact that ammonia concentrations are based on a number of factors (e.g., stocking density, water retention time, and time of feeding).

As an example of changes in effluent quality during cleaning, Kendra (1991) examined effluent quality during cleaning events at two hatcheries. At each hatchery, total suspended solids, total phosphorus, and BOD increased during cleaning. At one hatchery, TSS increased from 1 mg/L to 88 mg/L and total phosphorus increased from 0.22 mg P/L to 4.0 mg P/L. BOD increased from 3 mg/L to 32 mg/L at one facility, and at the other from 4 mg/L to 12 mg/L.

Boardman et al. (1998) conducted a study after surveys conducted in 1995 and 1996 by the Virginia Department of Environmental Quality (VDEQ), 2002, revealed that the benthic aquatic life of receiving waters was adversely affected by discharges from several freshwater trout farms. Three trout farms in Virginia were selected to represent fish farms throughout the state. This study was part of a

---

[2]Solids that are captured in quiescent zones or other in-process settling that occurs at flow-through system facilities are periodically cleaned out of the production units (i.e., quiescent zones, tanks, or raceways) to maintain optimal water quality in the process water. Accumulated solids, which can be about 60 to 70 percent of the total volume at a facility, are swept or vacuumed from the production units and conveyed to settling basins for treatment. The duration of the cleaning events range from a few minutes to about ½ hour or longer, depending on the size of the area being cleaned. The frequency of the cleaning events also varies based on the volume of solids that accumulate over time.

larger project to identify practical treatment options that would improve water quality both within the facilities and in their discharges to receiving streams.

After initial sampling and documentation of facility practices, researchers and representatives from VDEQ discovered that although pollutants from the farms fell under permit regulation limits, adverse effects were still being observed in receiving waters. Each of the farms was monitored from September 1997 through April 1998, and water samples were measured for dissolved oxygen (DO), temperature, pH, settleable solids (SS), TSS, total Kjeldahl nitrogen (TKN), total ammonia nitrogen (TAN), 5-day biochemical oxygen demand (BOD_5), and dissolved organic carbon (DOC).

Sampling and monitoring at all three sites revealed that little change in water quality between influents and effluents occurred during normal conditions at each facility (Table 7-1). The average concentrations of each regulated parameter (DO, BOD_5, TSS, SS, and AN) were below their regulatory limit at each facility; however, raceway water quality declined during heavy facility activity like feeding, harvesting, and cleaning. During these activities, fish swimming rapidly or employees walking in the water would stir up solids that had settled to the bottom. During a 5-day intensive study, high TSS values were correlated with feeding events. TKN and ortho-phosphate (OP) concentrations also increased during feeding and harvesting activities. Overall, most samples taken during this study had relatively low solids concentrations, but high flows through these facilities increased the total mass loadings.

Table 7-2 describes the water quality data for two flow-through systems sampled as part of EPA's data collection efforts at CAAP facilities. These results are comparable to those presented above. For both facilities there was little change between the influent and treated production effluent concentrations. However, pollutant concentrations in Off-Line Settling Business (OLSB) effluent was much higher than both influent and unit discharge waste effluent concentrations, and the OLSB flow rates were about one percent of the treated production unit discharge (Table 7-2).

7.2.2.3 Recirculating Systems

Recirculating systems have internal water treatment components that process water continuously to remove waste and maintain adequate water quality. Overall, recirculating systems produce a lower volume of effluent than flow-through systems. The effluent from recirculating systems usually has a relatively high solids concentration in the form of sludge. The sludge is then processed into two streams—a more concentrated sludge and a less concentrated effluent (Chen et al., 2002). Once solids are removed from the system, sludge management is usually the focus of effluent treatment in recirculating systems.

In a study describing the waste treatment system for a large recirculating research facility in North Carolina, Chen et al. (2002) characterize effluent at various points in the system (Table 7-3). Approximately 40% of the solid waste produced by this particular facility is collected in the sludge collector and composted. The remaining 60% of the solids are treated with two serial primary settlers (septic tanks) and then a polishing pond (receiving pond). Table 7-4 describes the water quality data for one recirculating system sampled as part of EPA's data collection efforts at CAAP facilities.

Table 7-1
Water Quality Data for Three Trout Farms in Virginia

Parameter	FARM A			FARM B			FARM C		
	Inlet	Within Farm	Outlet	Inlet	Within Farm	Outlet	Inlet	Within Farm	Outlet
Flow (mgd)	1.03–1.5 4[a] (1.18)[b]			4.26–9.43 (6.39)			9.74–10.99 (10.54)		
BOD$_5$ (mg/L)	0–1.2 (0.7)	0.5–3.9 (1.5)	0.96–1.9 (1.3)	0–1.4 (0.5)	0.3–7.2 (2.1)	0.6–2.4 (1.2)	0–2.0 (1.1)	0.4–7.5 (2.5)	0.5–1.8 (1.3)
DO (mg/L)	9.2–14.2 (10.6)	3.2–13.3 (7.0)	5.7–9.5 (8.5)	8.2–11.5 (10.5)	5.8–10.8 (8.6)	6.8–9.6 (7.9)	9.4–10.6 (10.5)	4.8–9.7 (7.6)	7.2–9.4 (8.1)
pH (SU)	7.1–7.4 (7.3)	7.0–7.4 (7.2)	7.3–7.8 (7.5)	7.3–7.6 (7.5)	7.2–7.6 (7.4)	6.9	7.3	7.1–7.6 (7.3)	7.8
Temp (°C)	10.5–13 (12.2)	11.5–15 (13)	11–15.5 (12.9)	6–12.5 (9.7)	6–14 (9.1)	5–16.5 (11.4)	8.5–13.5 (10.5)	8–14 (11.0)	8.5–14 (10.4)
TSS (mg/L)	0–1.1 (0.2)	0–30.4 (3.9)	0.8–6 (3.2)	0–1.8 (0.5)	0–43.7 (5.3)	1.5–7.5 (3.9)	0–1.5 (0.3)	0–28 (7.1)	4.1–62 (6.1)[c]
SS (mg/L)	ND		0–0.04 (0.02)	ND		0.01–0.08 (0.04)	ND		0.04–0.08 (0.07)
NH$_3$-N (mg/L)	0.6	0.2–1.1 (0.5)	0.5–0.6 (0.6)	0.2	0.06–1.1 (0.5)	0.45	0.03	0.03–2.2 (0.4)	0.02–0.17 (0.1)
DOC (mg/L)	0.93–4.11 (2.1)	0.9–7.9 (2.9)	1.5–2.4 (1.9)	0.91–2.56 (1.6)	1.2–8.1 (2.7)	1.2–3.1 (1.9)	1.1–2.7 (2.0)	1.1–11.1 (2.4)	1.5–3.8 (2.3)

[a] When available the range of values has been reported
[b] The average is indicated using italics.
[c] Two outliers were discarded for calculation of mean.
ND: Non-detect

Source: Boardman et al., 1998.

Table 7-2
Flow-through Sampling Data Table

Parameter	Facility A			Facility B		
	Inlet	OLSB Effluent	Bulk Water Discharge	Inlet	OLSB Effluent	Final Effluent
Flow (mgd)	192.4	0.914	91.4	2.481–2.777	0.017	2.481–2.777
BOD (mg/L)	ND (4)[a]	56.0–185.0[b] *(125.70)*[c]	3.50–4.20 *(3.85)*	ND (2)	13	ND (2)
pH (SU)	7.98–8.14 *(8.05)*	6.11–6.58 *(6.43)*	7.50–7.83 *(7.72)*	7.73–8.06 *(7.93)*	7.27	7.93–8.19 *(8.03)*
TSS (mg/L)	ND (4)	44.0–78.0 *(63.0)*	ND (4)	ND (4)	38	ND (4)
TP (mg/L)	0.7–0.25 *(0.14)*	8.32–11.10 *(9.81)*	0.15–0.25 *(0.21)*	0.02–0.03 *(0.03)*	0.36	0.03–0.07 *(0.05)*

[a] ND: Non-detect, the minimum level is listed in parenthesis.
[b] When available the range of values has been reported.
[c] The average is indicated using italics.
Source: USEPA sampling data. (Tetra Tech, 2002a)

Table 7-3
Water Quality Characteristics of Effluent at Various Points in the Waste Treatment System of Recirculating Aquaculture Systems at the North Carolina State University Fish Barn[a]

Parameter	Primary settling 1 inflow	Primary settling 2 inflow	Septic tank 2 outflow	Receiving pond effluent
COD (mg/L)	1043	690	409	153
TSS (mg/L)	752	364	205	44
TS (%)	0.22	0.18	0.16	0.11
NH_3-N (mg/L)	2.96	2.42	3.42	0.12
NO_2-N (mg/L)	5.35	31.17	44	1.93
NO_3-N (mg/L)	109	78.5	36.4	8.2
TKN (mg/L)	50.3	47.5	37.7	8.94
TP (mg/L)	28.6	22.7	17.6	4.95
PO_4-P (mg/L)	5.98	11.5	12.2	3.68

[a] Results are from sampling conducted 4 wk after startup of the waste handling system. Flow from the system into the receiving pond for the sampling period was 15.5 m^3/d.
Source: Chen et al., 2002.

Table 7-4
Recirculating System Sampling Data

Parameter	Facility C	
	Inlet	Discharge
Flow (mgd)	0.22	0.22
BOD (mg/L)	ND (2)[a]	35.0–48.0[b] *(42.0)*[c]
pH (SU)	7.8	6.97–7.25 *(7.15)*
TSS (mg/L)	ND (4)	26.0–60.0 *(42.80)*
TP (mg/L)	ND (0.01)	8.58–10.50 *(9.32)*

[a] ND: Non-detect, the minimum level is listed in parenthesis.
[b] When available the range of values has been reported.
[c] The average is indicated using italics.
Source: EPA sampling data. (Tetra Tech, 2001b)

7.2.2.4 Net Pen Systems

Although net pen systems do not generate a waste stream like other production systems, they do have a continuous, diluted discharge because of the tides and currents that provide a continual supply of high-quality water to flush wastes out of the system. In summarizing much of the 'Brooks' monitoring data in Puget Sound, Nash (2001) indicated that statistically significant increases in soluble (i.e., water column) nitrogen have been detected at salmon farms in Puget Sound, albeit infrequently, with no statistically significant increases 30 m downstream. Nash (2001) indicated that the maximum un-ionized ammonia levels were 0.0004 mg/L in comparison to a 4-day chronic water quality criterion of 0.035 mg/L (at a pH of 8 and 15°C). Nash also reported that in Puget Sound, dissolved (water column) inorganic nitrogen (DIN) ranged from 0.3 to 1.9 mg/L, while the maximum DIN increase due salmon farms was 0.09 mg/L.

Strain et al. (1995) estimated the nitrogen and phosphorus loadings in waters near Letang (New Brunswick, Canada) from 22 salmon farms by scaling the output from a fish growth model. Their estimates indicate nitrogen concentration increases from 0.03 to 0.07 mg/L and phosphorus increases from 0.0047 to 0.011 mg/L that are attributed to salmon aquaculture. Nitrogen, phosphorus, and BOD loadings from these salmon farms are the largest anthropogenic source of nitrogen, phosphorus, and BOD according to Strain et al. (1995).

7.2.2.5 Estimated Annual Loads for In-scope Flow-through and Recirculating Facilities

Estimated annual baseline loads for in-scope flow-through and recirculating facilities are presented in Figures 7-1 through 7-4. EPA used a facility-specific approach for estimating pollutant loads. EPA obtained detailed, facility-level information for a sample of potentially in-scope facilities

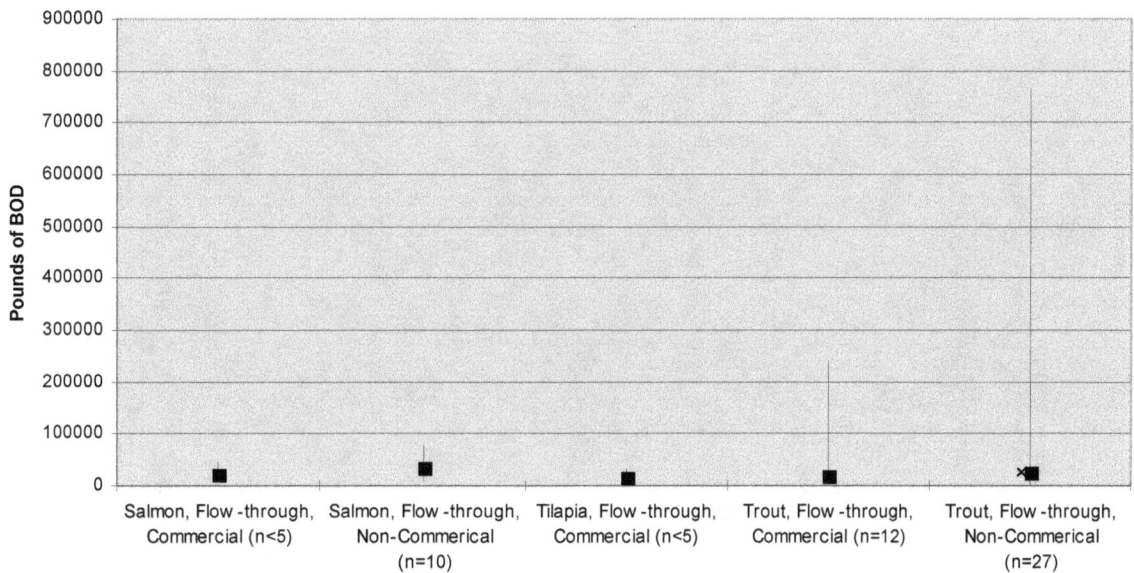

**Figure 7-1. Estimated Baseline Loads of BOD for In-scope Flow-through and
Recirculating Facilities.** The minimum value is indicated by the lowest point of the line, the
median by the square, and the maximum value by the highest point of the line. The number of
facilities on which the minimum, median, and maximum values are based is indicated in
parentheses under each group label.

Please see Section 7.2.2.5 and Chapter 10 of the Technical Development Document for more
information.

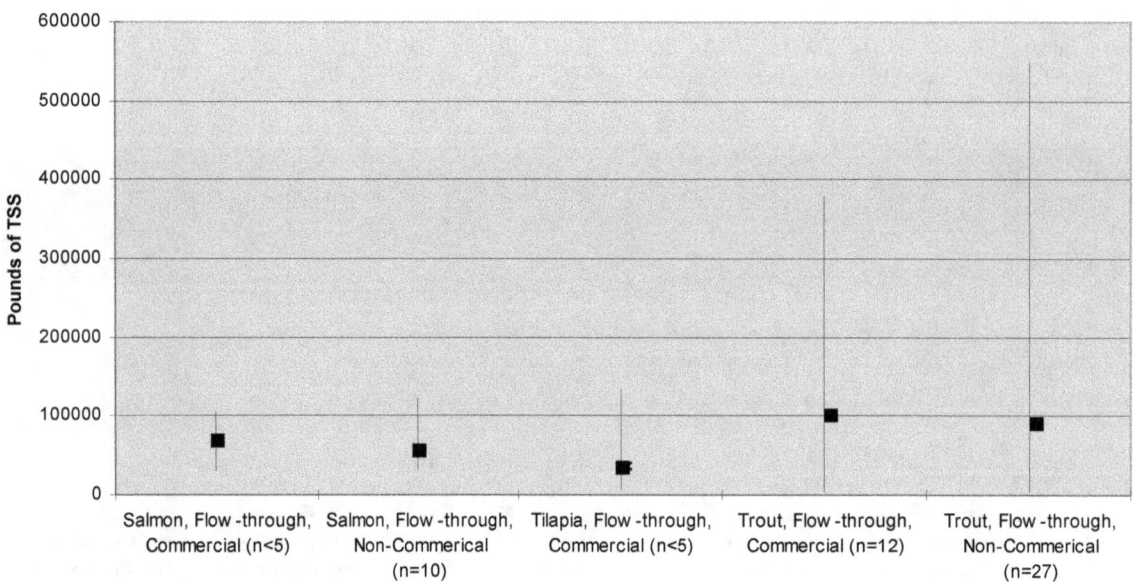

Figure 7-2. Estimated Baseline Loads of TSS for In-scope Flow-through and Recirculating Facilities. The minimum value is indicated by the lowest point of the line, the median by the square, and the maximum value by the highest point. The number of facilities on which the minimum, median, and maximum values are based is indicated in parentheses under each group label.

Please see Section 7.2.2.5 and Chapter 10 of the Technical Development Document for more information.

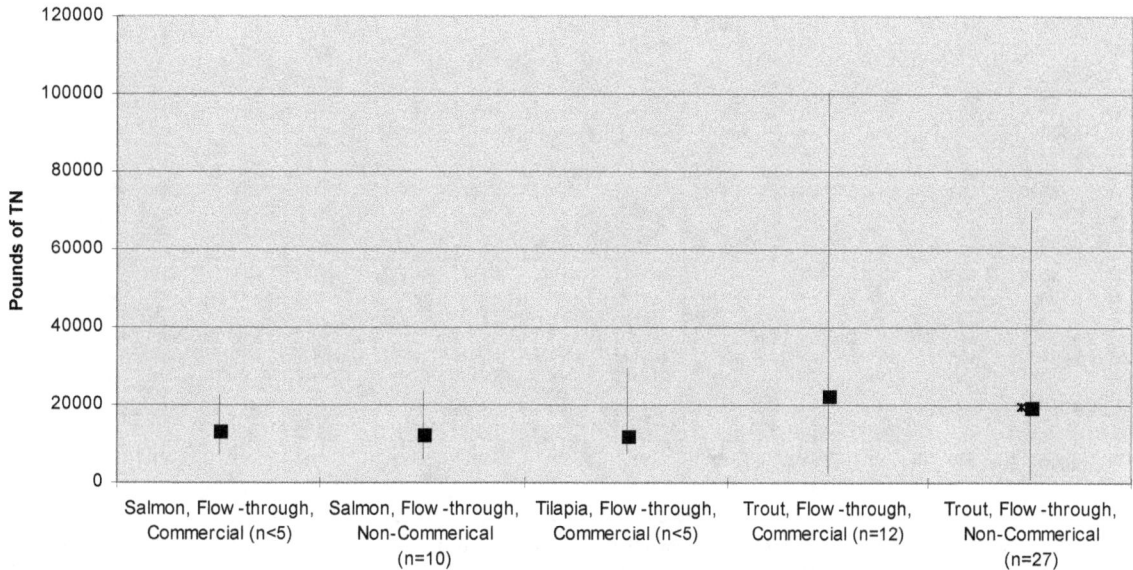

Figure 7-3. Estimated Baseline Loads of Total Nitrogen for In-scope Flow-through and Recirculating Facilities. The minimum value is indicated by the lowest point of the line, the median by the square, and the maximum value by the highest point of the line. The number of facilities on which the minimum, median, and maximum values are based is indicated in parentheses under each group label.

Please see Section 7.2.2.5 and Chapter 10 of the Technical Development Document for more information.

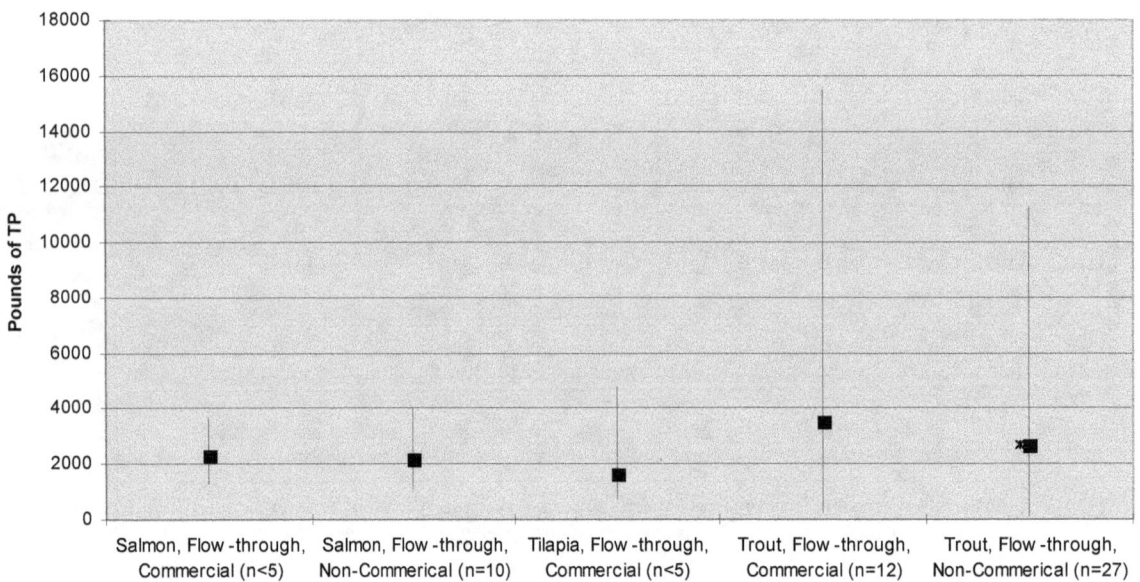

Figure 7-4. Estimated Baseline Loads of Total Phosphorus for In-scope Flow-through and Recirculating Facilities. The minimum value is indicated by the lowest point of the line, the median is represented by the square, and the maximum value is indicated by the highest point of the line. The number of facilities on which the minimum, median, and maximum values are based is indicated in parentheses under each group label.

Please see Section 7.2.2.5 and Chapter 10 of the Technical Development Document for more information.

through detailed AAP survey (USEPA, 2002a). EPA analyzed the detailed survey information, specifically information about feed inputs from which baseline loads for TSS, BOD, total nitrogen (TN), and TP could be estimated. Refer to Chapter 10 of the Technical Development Document for additional information.

7.2.3   Metals and Feed Additives/Contaminants

Metals may be present in CAAP wastewaters due to a variety of reasons. They may be used as feed additives, occur in sanitation products, or may result from deterioration of CAAP machinery and equipment. EPA has observed that many of the treatment systems used within the CAAP industry provide substantial reductions of most metals since most of the metals can be adequately controlled by controlling solids. Trace amounts of metals are added to feed in the form of mineral packs to ensure that the essential dietary nutrients are provided. Examples of metals added as feed supplements include copper, zinc, manganese, iron (Snowdon, 2003). Estimated baseline loads of metals and other feed additive/contaminants for in-scope facilities are summarized in Hochheimer et al., 2004. These loads were estimated as a function of TSS loads, using data obtained from samples collected by EPA during three sampling episodes (see Tetra Tech, 2001a, 2001b, and 2002a for detailed information on these sampling episodes) performed for the proposed rule. For this analysis, EPA set the analyte concentration in samples in which the analyte was not detected equal to one-half the detection limit of the analytical method used. From the sampling data, EPA calculated net TSS and metals concentrations at different points in the hatcheries. EPA then calculated metal-to-TSS ratios (in mg of metal per kg of TSS), based on net concentrations calculated above, and removed negative and zero ratios from the sample. Finally, basic sample distribution statistics were calculated to derive the relationship between TSS and each metal. Refer to Chapter 10 of the Technical Development Document for more information (USEPA, 2004).

Two substances, astaxanthin and canthaxanthin, are added to feed of farmed fish to improve consistent coloring of fish tissue. Astaxanthin and canthaxanthin have been widely used in northwestern Europe and North America, particularly for the artificial coloration of the flesh of salmonids during the later stages of grow-out operations (GESAMP, 1997). Two organisms, phaffia yeast (*Phaffia rhodozyma*) and haematococcus algae meal (dried *Haematococcus pluvialis*), produce astaxanthin and are certified by the FDA as approved color additives in fish feed (21 CFR 73.355 and 73.75). Pure astaxanthin, phaffia yeast, or haematoccus algae meal can be added to fish feed to induce the desired coloration in fish. *Phaffia rhodozyma* yeast naturally synthesizes astaxanthin during fermentation.[3] EPA has not attempted to quantify potential loads for these additives.

---

[3]The European Commission Scientific Committee for Animal Nutrition (SCAN) examined the use of astaxanthin-rich *Phaffia rhodozyma* yeast in salmon and trout feed. In this report, the Committee noted that although safety aspects for the yeast were satisfactorily demonstrated, questions on effects of the active ingredient, astaxanthin, on the environment, remained an open question, despite assertions by the company that there was no need to study excreted residues because astaxanthin is present in nature and the product is a true (dead) yeast and as such an accepted feed ingredient (SCAN, 2002a). Regarding canthaxanthin, SCAN recommended in a 2002 opinion that the maximum permitted concentrations of canthaxanthin in feed be reviewed to ensure consumer safety (SCAN, 2002b; the committee was not asked to address the potential impact of canthaxanthin use on the environment and as a result there is no reference to this aspect of the assessment in the 2002 opinion). In 2003, the European Union amended permitted canthaxanthin levels in feed for salmonids and other animals in order to provide greater protection for consumers' health (Commission Directive 2003/7/EC of 24 January 2003, as reported in the Official Journal of the European Commission, January 25, 2003).

Several efforts have been made to evaluate aquaculture feed contaminants. As part of a recent investigation of organic contaminants in farmed salmon, Hites et al. (2004) analyzed thirteen samples of commercial salmon feed from Europe and North and South America for organochlorine contaminants. The authors found that while concentrations in feed were quite variable, they observed that concentrations in feed purchased from Europe were significantly higher than those in feed purchased from North and South America, possibly reflecting lower contaminant concentrations in forage fish from the coastal waters of North and South America as compared with those from the industrialized waters of Europe's North Atlantic (Hites et al., 2004). The U.S. Geological Survey (USGS) has sampled and is analyzing the occurrence of metal and organochlorine contaminant residues in commercial feeds purchased by the U.S. Fish and Wildlife Service hatcheries as a result of nutritional problems that were observed at some FWS hatcheries (USGS, n.d.). The results were being analyzed in late 2003 (J. Bayer, USGS, personal communication with L. McGuire, U.S. EPA, December, 2003 (McGuire, 2004)). EPA developed crude estimates of polychlorinated biphenyls (PCBs) in baseline loads for in-scope facilities, as summarized in Hochheimer et al., 2004.

7.2.4   Other Contributions and Releases

Maintenance of the physical plant of aquaculture facilities can generate organic materials that may contribute to water quality degradation (NOAA, 1999). For example, the activity of cleaning fouling organisms from net pens can contribute solids, BOD, and nutrients, although these inputs are generally produced only over a short period of time. Cleaning algae from flow-through raceway walls and bottoms similarly generates pollutants in effluent.

Cultured fish themselves may be lost from facilities because of decomposition of carcasses or scavenging by birds, mammals, and fish (Nash, 2001). Leakage may occur from small holes in net-pens, during handling, or during transfer of fish to another pen, and fish may also be lost as a result of operator error, predation, storms, accidents, and vandalism. One author writes:

> "It is widely known among commercial fish culturists that when fishes are held within nets in a body of water, a certain portion of fish assumed to be in cages disappears...this unexplained loss of fishes has been recognized for decades....Even today, commercial fish culturists continue to lose important numbers of salmonid fishes from cages in salt water, estimated to range from 10% to as much as 30%....Fish disappear even when there are no tears in the netting, the cages are covered, and daily inspections of cages are made...this loss can have economic importance - not only because of lost fish ("shrinkage") but also because food provided for these "phantom" fish often falls through the bottom netting and is wasted, such that assumed feeding rates and food conversions thus are both inflated." (Moring, 1989)

Based on his study of losses from net-pen facilities in Puget Sound, the author attributed losses to decomposition of carcasses, particularly during disease outbreaks; scavenging by birds, mammals, and fishes, and to a lesser extent, escapes, when cage netting remains intact (Moring, 1989). Various estimates of numbers of escaped fish from some net-pen facilities in the U.S. have been noted elsewhere (e.g., USEPA, 2002b).

EPA did not attempt to quantify other contributions and releases such as those described above from facilities in the scope of EPA's final regulation.

7.2.5   Drugs and Pesticides

By providing food and oxygen, AAP facilities can produce fish and other aquatic animals in greater numbers than natural conditions would allow. This means that system management is important to ensure that the animals do not become overly stressed, making them more vulnerable to disease outbreaks. When diseases do occur, facilities might be able to treat their populations with drugs.

FDA/Center for Veterinarian Medicine (CVM) regulates animal drugs under the Federal Food, Drug, and Cosmetic Act (FFD&CA). Operators producing aquatic animals that are being produced for human consumption must comply with requirements established by the U.S. Food and Drug Administration (FDA) with respect to the drugs that can be used to treat their animals, the dose that can be used, and the withdrawal period that must be achieved before the animals can be harvested. Four categories of drugs are used in aquaculture: (1) six commercial drugs currently approved for specific species, specific diseases, and at specific doses or concentrations; (2) investigational new animal drugs which are used under controlled conditions under an Investigational New Animal Drug (INAD) application; (3) the extralabel use of FDA-approved drugs under the provisions of the Animal Medicinal Drug Use Clarification Act of 1994 (AMDUCA); and (4) drugs designated by FDA as low regulatory priority. Pesticides are also used to control animal parasites and aquatic plants at CAAP facilities

FDA/CVM approves new animal drugs based on scientific data provided by the drug sponsor. These data include environmental safety data that are used in an environmental risk assessment for the drug (Eirkson et al., 2000). Approved drugs have already been screened by the FDA to ensure that they do not cause significant adverse public health or environmental impacts when used in accordance with label instructions. See Section 7.3.3 for more information on FDA/CVM's environmental review process.

Currently, there are only six approved drugs for AAP species consumed by humans:

- Chorionic gonadotropin (Chorulon®)
- Oxytetracycline (Terramycin®)
- Sulfadimethoxine, ormetoprim (Romet-30®)
- Tricane methanesulfonate (Finquel® and Tricaine-S)
- Formalin (Formalin-F®, Paracide-F® and Parasite-S®)
- Sulfamerazine®

Investigational new animal drugs (INADs) are those drugs for which FDA has authorized use on a case-by-case basis to allow a way of gathering data for the approval process (21 USC 3606(j)). Quantities and conditions of use are specified. FDA, however, sometimes relies on the NPDES permitting process to establish limitations on pollutant discharges to prevent environmental harm.

Extralabel drug use is restricted to use of FDA-approved animal and approved human drugs by or on the lawful order of a licensed veterinarian within the context of a valid veterinarian-client-patient relationship. Specific conditions governing the extralabel use of drugs are established in 21 CFR Part 530. Specific conditions and provisions in 21 CFR Part 530 include those relating to compounding of approved new animal and approved human drugs, extralabel use in food-producing and non-food producing animals, safe levels and analytical methods, and specific drugs, families of drugs, and substances prohibited for extralabel use in animals. As stated in 21 CFR Part 530, extralabel use is limited to treatment modalities when the health of an animal is threatened or suffering or death may result from failure to treat. Extralabel uses that are not permitted include uses that result in any residue which may present a risk to public health and uses that result in any residue above an established safe level, safe concentration or tolerance. Additionally, AMDUCA prohibits the use of an FDA approved drug in or on any animal feed. See 21 CFR Part 530 for more detail on extralabel use conditions and limitations.

Unapproved new animal drugs are sometimes used in discrete cases where the FDA exercises its regulatory discretion. In determining whether a compound can be used without a New Animal Drug Application (NADA), FDA considers human food safety (if for use in food animals/fish), user safety and any other impacts of the unapproved use. Regulatory discretion does not constitute an approval by the Agency nor an affirmation of their safety and effectiveness. The FDA is unlikely to object to the use of any of these drugs if the substances are used under specific indications, at the indicated levels, and according to good management practices. In addition, the product should be of an appropriate grade for use in food animals (FDA, 1997). The user of any of the low regulatory compounds is responsible for meeting all local, state and federal environmental requirements.

The FDA does not require labeling for low-priority use for chemicals that are commonly used for non-drug purposes even if the manufacturer or distributor promotes the chemical for the permitted low-priority use. However, a chemical that has significant animal or human drug uses in addition to the low-priority aquaculture use must be labeled for the low-priority uses if the manufacturer or distributor uses promotion or other means to establish the intended low-priority use for the product. Additional labeling requirements are available from the FDA (FDA, 1997).

Pesticides may also be used to control animal parasites and aquatic plants and may be present in wastewaters from CAAP facilities.

Aquatic animal production facilities use a number of drugs and pesticides for a variety of reasons. Refer to Table B-1 in Appendix B for more specific information about drugs and pesticides used at aquatic animal production facilities and their generally reported uses.

MacMillan (2003) estimates that between 50,000 and 70,000 pounds of antibiotic active ingredient are sold each year for use in the aquaculture industry (0.3-0.4% of all antibiotics used in animal agriculture). For a summary of the total amount of drugs and pesticides used during 2001 by aquatic animal production facilities that completed detailed surveys, refer to Hochheimer and Meehan, 2004a.

7.2.6   Pathogens

CAAP facilities are not considered a source of pathogens that adversely affect human health. CAAP facilities culture cold-blooded animals (fish, crustaceans, mollusks, etc.) that are unlikely to harbor or foster such pathogens (MacMillan et al., 2002). EPA sampling data also supports this assertion (Tetra Tech, 2001a, 2001b; Tetra Tech, 2002a). Although it is possible for CAAP facilities to become contaminated with human pathogens (e.g., by contamination of facility or source waters by wastes from warm-blooded animals) and, as a result, become a source of human pathogens, this is not considered a substantial risk in the United States (MacMillan et al., 2002).

7.3   IMPACTS OF CAAP INDUSTRY DISCHARGES

7.3.1   Impacts from Solids, Nutrients, BOD, Metals, and Feed Contaminants

As described in more detail in Section 7.2, CAAP facility effluents can contribute nutrients (nitrogen and phosphorus), suspended solids, and BOD to receiving waters. Impacts associated with CAAP facility discharges include stimulation of algal and aquatic vascular plant growth, sediment oxygen demand, and a variety of chemical/biochemical processes that consume oxygen. The following sections first describe general aquatic ecosystem effects of discharges of solids, nutrients and BOD on aquatic ecosystems and then summarize recent literature reporting observations of AAP and/or CAAP facility discharges on aquatic ecosystems.

7.3.1.1   General Aquatic Ecosystem Effects

Solids (i.e., total suspended solids or TSS) are discharged from CAAP facilities both as suspended and settleable forms, primarily from feces and uneaten feed. Since TSS in effluents from CAAP facilities contains a high percentage of organic content, these solids can contribute to eutrophication and oxygen depletion (when microorganisms decompose the organic matter and consume oxygen). Suspended solids can also degrade aquatic ecosystems by increasing turbidity and reducing the depth to which sunlight can penetrate, which may decrease photosynthetic activity and growth of aquatic vascular plants and algae. Increased suspended solids can also increase the temperature of surface water because the particles may absorb heat from sunlight. Excess TSS can also cause a shift toward more sediment-tolerant species, carry nutrients and metals, and adversely affect aquatic insects that are at the base of the food chain (Schueler and Holland, 2000). As sediment settles, it can smother fish eggs and bottom-dwelling organisms, interrupt the reproduction of aquatic species, and destroy habitat for benthic organisms (USEPA, 2000). Suspended solids have been associated with effects on fish including reduced food consumption by certain life-stages of species (Breitburg, 1988; Redding et al., 1987; Gregory and Northcote, 1993).

Nutrients in the CAAP discharge can stimulate the growth of algae and other aquatic plants. Although algae and aquatic plants produce oxygen as a by-product of photosynthesis, they are net consumers of oxygen during periods of respiration when photosynthesis is not occurring due to absent or very limited sunlight. Many of the organic solids discharged from CAAP facilities settle rapidly and decompose at the sediment-water interface, which is termed sediment oxygen demand (Schueler and Holland, 2000). As discussed above, solids may lead to increased water temperatures, which ultimately decreases oxygen (warmer water has lower oxygen saturation levels). Other chemical and biochemical reactions, such as nitrification, also consume oxygen. The combination of eutrophication, plant growth, sediment oxygen demand, warming, and chemical or biochemical reactions may lead to changes in local or downstream dissolved oxygen. Often the net change is a lowering of oxygen levels available for aquatic and benthic organisms. Dissolved oxygen is essential to the metabolism of all strict aerobic aquatic organisms and its distribution in aquatic environments affects chemical, biological, and ecological processes (Wetzel, 1983).

Nitrogen at CAAP facilities can come from several sources. The largest contributor of nitrogen in effluents from CAAP systems comes from fish feed and feces (Avault, 1996). In CAAP facilities, nitrogen is mainly discharged as ammonia, nitrate, and organic nitrogen. Organic nitrogen decomposes in aquatic environments into ammonia and nitrate. Ammonia can be directly toxic to aquatic life, affecting hatching and growth rates of fish. However, ammonia is not usually found at toxic levels in CAAP discharges.

CAAP facilities release phosphorus in both the solid and dissolved forms. The dissolved form, generally as orthophosphate, is more readily available to plants and bacteria, which require phosphorus for their nutrition (Henry and Heinke, 1996). Excessive amounts of orthophosphate in the aquatic environment increase algae and aquatic plant growth, especially in freshwater environments where phosphorus is more likely to be a limiting nutrient. Although the solid form of phosphorus is generally unavailable, depending on the environmental conditions (e.g., availability of oxygen), some phosphorus may be slowly released from the solid form.

CAAP discharges to receiving waters of feed contaminants include metals and organochlorines and are discussed in Section 7.2.3. There is limited evidence that these contaminants adversely affect aquatic ecosystems in the United States under current practices. In an examination of the potential for heavy metal accumulation beneath net-pen farms in the Pacific Northwest, sediment concentrations of zinc, an essential trace element added to salmon feeds as part of the mineral supplement, were found to be typically increased near salmon farms (Nash, 2003). However, environmental factors (e.g., sediment sulfide concentrations), natural attenuation, advances in feed formulations, and existing net-pen benthic monitoring requirements are asserted to mitigate the potential for toxic levels to occur (Nash, 2003; Brooks and Mahnken, 2003a and 2003b). Although EPA is aware of recent interest in contaminants found in salmonid feed and farmed salmon (e.g., USGS, n.d.; Hites et al., 2004), EPA is aware of no peer-reviewed studies of the effects of releases of organochlorine contaminants in aquaculture facility wastes to receiving waters and limited evidence that such releases may pose an ecological risk. Easton et al. (2002) cite unpublished 1987 data from British Columbia indicating that benthic organisms around net-pen facilities contained elevated levels of polychlorinated biphenyls (PCBs) originating from salmon feed but no indication that these levels posed an ecological risk was provided. Internal documents prepared by the Pennsylvania Department of Environmental Protection also report elevated levels of PCBs in a small number of sediment, fish, and invertebrate samples from receiving water environments at several Pennsylvania hatcheries (McGuire, 2004).

Discharges of approved drugs and pesticides and other treatments used at aquaculture facilities may also impact aquatic ecosystems. For example, releases of copper compounds, used as antifoulants in raceways, tanks, and on net-pens, may lead to receiving water effects including changes to dissolved oxygen levels as algae die from exposure (Cornell, 1998). Nash (2003) concluded that potential risk from elevated sediment copper concentrations from marine net-pen anti-fouling compounds could be significantly reduced both by environmental factors (e.g., sediment sulfide concentrations, natural attenuation processes), as well as management practices such as washing nets at upland facilities and properly disposing of the waste in an approved landfill. Section 7.3.3 discusses in more detail literature regarding environmental effects of approved drugs.

7.3.1.2 Recent Literature

The previous section describes in a general sense the role that excess solids, nutrients, BOD, and feed contaminants could play in aquatic ecosystems. Studies discussed in this section include several site-specific studies related to aquatic ecosystem effects of effluent discharges from aquaculture facilities. Other literature describing aquatic ecosystem effects of facility discharges has been described elsewhere (e.g., work reported in RAC, 1998; USEPA, 2002b, Appendix E: "Literature Review for AAP Impacts on Water Quality").

Loch et al. (1996) examined the effects of three large trout flow-through facilities in North Carolina on macroinvertebrate species diversity. Their data showed that species richness was significantly decreased below the outfalls of the facilities. Samples did show that richness did increase further downstream. These data indicate that effluents did reduce water quality, even at 1.5 km further downstream, although to a lesser extent. The authors noted that impacts were seasonal, and that water quality and taxa richness improved during the winter. The authors also noted that sewage fungus (which they defined as a community of organisms that consist mainly of bacteria and ciliated protozoans and is the product of concentrated organic matter) "was present in great abundance at Site 2 of each trout farm."

In contrast, Fries and Bowles (2002) examined aquatic impacts associated with a large CAAP facility located on the San Marcos River in Texas, which is designated by the Texas National Resource Conservation Commission as exceptional for aquatic life and recreation. On average, this CAAP facility produces four million largemouth bass fingerlings, one million channel catfish fingerlings, 12,000 kg live forage for captive broodstock, and 67,000 rainbow trout (winter only) each year. Based on the data covering a period from October 1996 to July 1998, the authors concluded that "the hatchery effluent did not substantially affect downstream water quality and benthic communities, despite the relatively high total suspended solids and chlorophyll-a levels in the effluent." The authors noted "...that sportfish hatchery operations can have negligible effects on receiving waters, even in environmentally sensitive systems."

In the 1970s, Big Platte Lake in Michigan, which is fed by the Platte River, was experiencing periods of calcium carbonate formation that were reducing lake transparency (also called "whiting"), as well as other symptoms of eutrophication including reduced macroinvertebrate communities and disappearance of sensitive vegetation. Because the watershed is mostly undeveloped, a possible explanation of these changes in lake conditions was phosphorus loadings from nonpoint sources, effluents from the Platte River State Fish Hatchery, salmon smolts dying in outmigration, and returning adult salmon deaths in the river. It was estimated that the hatchery was contributing approximately 33% of the phosphorus load into the lake in the late 1970s (Whelan, 1999). In its 1980 NPDES permit, the hatchery was required to take steps to reduce phosphorus loads in its effluent. However, subsequent court cases found that significant changes in facility operation would be required to mitigate the impairment of Big Platte Lake. Beginning in 1998, the hatchery took further actions to improve lake conditions. The hatchery's 1988 NPDES permit restricted water use to 166 million liters per day, with a maximum discharge of 200 kg of phosphorus per year, and TSS limits of 1,000 kg/day. Through the use of low phosphorus fish food, improvements in waste removal, deepening of treatment ponds, and changes in fish migration, the hatchery now contributes only 5% of the annual phosphorus loading to the lake. Maximum transparency in the lake has increased from an average of 3.5 meters to 5 meters or greater. Severe whiting events continue to occur during the summer months, although these loss of transparency problems are less frequent since 1988. Studies and renovations of the hatchery are estimated to further improve water conditions in the future (Whelan, 1999).

Memoranda, correspondence, and discussion with staff of the South Central Region of the Pennsylvania Department of Environmental Protection (PA DEP) indicate environmental impacts at several CAAP facilities (200,000 to 400,000 lbs annual production) in Pennsylvania. PA DEP provided data and reports documenting adverse impacts of hatchery effluents in receiving spring-fed streams. The materials described observations and/or concerns including those about discharges of carbonaceous BOD and TSS and other pollutants, and results of aquatic biological surveys showing adverse impacts in hatchery receiving waters. While recognizing unique characteristics of these hatcheries (all located on limestone spring creeks and all capture most, if not all, of the streamflow) and seasonality of these impacts, staff biologists were concerned about adverse environmental impacts observed at several sites (Botts, 1999; Embeck, 2000; Botts, 2001; McGuire, 2003).

EPA performed a review of literature to document reported water quality impacts from net pen facilities (Mosso et al., 2003). Literature showed that organic enrichment may result from uneaten feed or feces that accumulate on sediment below and near the perimeter of net pens. McGhie et al. (2000) showed that the rate of accumulation is affected by the amount of uneaten feed and feces from the production facility as well as the amount of material transported away from the site largely as a result of water current velocities. Effects of organic enrichment include changes in benthic communities such as recruitment of organic carbon tolerant species and diminution of organic carbon sensitive species. These changes may result in reduced diversity and abundance or organisms (Nash, 2001; Findlay et al., 1995; McGhie et al., 2000; La Rosa et al., 2001). In addition, organic loading to the sediment might exceed existing benthos capacity and might become anoxic. Anoxia can lead to further changes in benthic communities as well as to sediment chemistry changes, including increased sulfide concentrations and decreased redox potential, which are common at net pen facilities. Literature examined shows that the nearfield impacts to benthic communities are common within 100 meters of the net-pen perimeter. Many net pen operators routinely fallow net pen sites on a regular basis (Bron et al., 1993) primarily for disease and parasite control, but also to reduce benthic impacts. For example, many Maine net pen operators raise single year classes at a site and fallow the site for about 30 to 90 days after harvest, depending on temperature, currents, and benthic conditions (Tetra Tech, 2002b and 2002d).

In addition to literature described above and elsewhere, several Total Maximum Daily Load (TMDL) reports describe aquaculture facilities' contributions to pollutant loads in specific watersheds. The following paragraphs describe several such TMDL documents. The brief descriptions below are not meant to imply that TMDLs involving aquaculture facilities are prevalent, but rather only to illustrate that several have been developed, and to illustrate the types of pollutants that are addressed.

In 2002, the Virginia Department of Environmental Quality's Virginia Water Resources Research Center submitted a report, *Benthic TMDL Reports for Six Impaired Stream Segments in the Potomac-Shenandoah and James River Basins*. This document reports on a Total Maximum Daily Load (TMDL) calculation performed for six impaired stream segments in Virginia. These stream segments were listed as impaired on EPA's 1998 303(d) report following benthic macroinvertebrate surveys. Critical stressors to these stream segments were identified, and the report concludes that aquaculture effluents were confirmed as the primary source of the organic solids that impaired these short segments (0.02 to 0.8 miles). The aquaculture facilities constituted from 86.2 percent (11,481 pounds per year out of 13,325 total pounds per year for the particular stream) to 99.6 percent (4,438 pounds per year out of 4,455 total pounds per year for the particular stream) of the organic solids loading in these sections of primarily first-order, spring-fed streams. To put these loads into perspective, the organic load (defined as 60 percent of the measured TSS load from a facility) to the different streams ranged from 1,823 pounds per year (94.9 percent of the total load in the particular stream) to 72,477 pounds per year (99.4 percent of the total load in the particular stream) (VDEQ, 2002).

A number of TMDLs have been developed to address water quality concerns associated with pollutant loads from sources including aquaculture in the Snake River region Idaho. The Middle Snake River, Idaho, is a 150 km stretch of the Snake River that has been transformed from a free-flowing river to one with multiple impoundments, flow diversions, and increased pollutant loadings. These changes have led to significant alterations to river habitat, loss of native macroinvertebrate species, extirpation of native fish species, expansion of pollution-tolerant organisms, and excessive growth of macrophytes and algae. According to EPA (2002), 80 private and State-owned aquaculture facilities operate under federal NPDES permits, and over 20 additional facilities have applied for permits, in the Middle Snake River. These facilities supply approximately 80% of the trout consumed in restaurants in the United States (USEPA, 2002c). TMDLs in various stages of completion which address loadings from many of the aquaculture facilities in this region include the Upper Snake Rock TMDL, the Billingsley Creek TMDL, and the Cascade Reservoir TMDL. Pollutants addressed in these TMDLs include total phosphorus total suspended solids.

In a TMDL for a small reservoir in Utah, aquaculture was identified as a significant contributor to an impaired water, resulting in a recommended load reduction of 15% (13.2% of the total load reduction recommended) from the hatchery discharging to the impaired reservoir (Utah DEQ, 2000). According to the TMDL report:

"Mantua Reservoir is a small reservoir located within the community of Mantua in east Box Elder County, Utah...Mantua Reservoir is highly productive (i.e., has a large amount of nutrients such as nitrogen and phosphorus), creating problems that include dense beds of aquatic plants, algal blooms, low dissolved oxygen (DO), and high pH. The high productivity is primarily due to the lake's shallowness and excess loading of nutrients from the watershed....The Mantua Fish Hatchery is the only permitted point source in the watershed....[and] is a significant contributor of nutrients to the Reservoir, adding an estimated 304.4 kg/Y TP (31% of total load)..."

7.3.2 Impacts from Other Releases

Other releases from facilities (discussed in Section 7.2.4) include materials related to maintenance activities, loss of fish via decomposition of carcasses, and escapes. In some cases, escaped cultured organisms may not be native to the receiving water and at certain levels may pose an environmental risk. Scientists and resource managers have recognized aquaculture operations as a potential source of concern with respect to non-native species issues (ADFG, 2002; Carlton, 2001; Goldburg et al., 2001; Naylor et al., 2001; Lackey, 1999; and Volpe et al., 2000). It is important to note, however, that many non-native fishes are introduced intentionally. For example, non-native sport fish species are a large and important component of a number of state recreational fishery programs. Horak (1995) reported that "[forty]-nine of 50 state recreational fishery programs use nonnative sport fish species, and some states are almost totally reliant on them to provide recreational fishing." This section does not address such intentional releases. In addition, scientists have also highlighted the need for careful assessment of potential environmental risks associated with the possible future use of genetically modified organisms in aquatic animal production (e.g., Hedrick, 2001; Reichardt, 2000; Howard et al., 2004).

Many states have developed requirements specific to potential escapes of non-native organisms from aquaculture facilities (see, for example, the tilapia discussion under Section 7.3.2.2) and/or have developed aquatic nuisance species (ANS) management plans to address non-natives in their state. ANS management plans identify goals or objectives for addressing ANS and strategic actions or tasks to accomplish the goals or objectives. For example, an objective might be to prevent the introduction of new ANS into state waters. A strategic action to accomplish this might be to identify those ANS that have the greatest potential to infest state aquatic resources. As part of this effort, states might identify existing and potential pathways that facilitate new ANS introductions. A task that might be used to accomplish the strategic action might be to develop a regional listing of ANS and evaluate the potential threat posed by these organisms to aquatic resources in the state. ANS management plans are available on the Aquatic Nuisance Species Task Force website at *http://www.anstaskforce.gov*.

The following sections describe general issues relating to effects of non-native aquatic organisms (Section 7.3.2.1) and specific discussions relating to non-native issues specifically related to aquaculture operations (Section 7.3.2.2).

7.3.2.1 General Aquatic Ecosystem Effects

Non-native aquatic organisms in North America can alter habitat, change trophic relationships, modify the use and availability of space, deteriorate gene pools, and introduce diseases. Non-native fish introduced to control vegetation, such as carp or tilapia, can destroy native vegetation. Destruction of exotic and native vegetation can result in bank erosion, degradation of fish nursery areas, and acceleration of eutrophication as nutrients are released from plants. Common carp (*Cyprinus carpio*) reduce vegetation by direct consumption and by uprooting as they dig through the substrate in search of food. Digging increases turbidity in the water (AFS, 1997; Kohler and Courtenay, n.d.). Non-native species may also cause complex and unpredictable changes in community trophic structure. Communities can be changed by explosive population increases of non-native fish or by predation of native species by introduced species (AFS, 1997). Spatial changes may result from overlap in the use of space by native and non-native fish, which may lead to competition if space is limited or of variable quality (AFS, 1997).

Genetic variation may be decreased through inbreeding by species being produced in a hatchery. If these species are introduced to new habitat, they may lack the genetic characteristics necessary to adapt or perform as predicted. There is also a possibility that native gene pools may be altered through hybridization from non-native species. However, hybridization events in open waters are rare (AFS, 1997; Kohler and Courtenay, n.d.). Finally, diseases caused by parasites, bacteria, and viruses may be transmitted into an environment by non-native species. For example, transfer of diseased non-native fish from Europe is believed to be responsible for introducing whirling disease in North America (Blazer and LaPatra, 2002).

7.3.2.2 Recent Literature

The following discussions of Atlantic salmon and tilapia illustrate the potential or actual role of aquatic animal production in releases of non-native species. These organisms are discussed here because they are known to be cultured at facilities such as those in the scope of the final CAAP Rule. In the case of Atlantic salmon, EPA received many comments regarding potential environmental impacts of farmed, non-native salmon escaping from net pens and is aware that these species are raised in marine net pens in both the Puget Sound and New England areas. Tilapia species are known to be raised at CAAP facilities in the scope of the final regulation, and again, EPA is aware of concerns that have been raised with the potential establishment of this group of non-North American species.

It should be noted that other aquaculturally raised organisms have been identified as a source of concern in some environments by resource managers and scientists from a non-native species perspective (e.g., carp, Asian oysters). For example, grass carp (*Ctenopharyngodon idella*) have spread rapidly in the last few decades from research projects, escapes from natural ponds and aquaculture pond facilities, legal and illegal interstate transport, releases by individuals and groups, stockings by Federal, State, and local government agencies, and natural dispersion from introduction sites (Pflieger, 1975; Lee et al., 1980; Dill and Cordone, 1997). Grass carp remove vegetation, which can result in the elimination of food, shelter, and spawning substrates for native fish (Taylor et al., 1984). Black carp (*Mylopharyngodon piceus*) provide a cheap means for controlling trematodes in catfish ponds, but they feed on many different mollusks when released to the environment. Silver carp (*Hypophthalmichthys molitrix*) were discovered in natural waters in 1980, "probably a result of escapes from fish hatcheries and other types of aquaculture facilities" (Freeze and Henderson, 1982, as cited in Fuller et al., 1999). Bighead carp (*Hypophthalmichthys nobilis*) first appeared in open waters (Ohio and Mississippi rivers) in the early 1980s, "likely as a result of escapes from aquaculture facilities (Jennings 1988, as cited in Fuller et al., 1999). Both carp have been identified as species of significant concern to aquatic resource managers (Schomack and Gray, 2002). Again, however, it is important to stress that carp are mainly raised in pond aquaculture systems, and that pond systems are not in the scope of EPA's final regulation.

Atlantic Salmon

Escapement of Atlantic salmon (*Salmo salar*) from net pens off the East and West Coasts of the United States and in British Columbia has been well documented. Potential concerns associated with Atlantic salmon escapes include possible impacts on wild salmon from disease, parasitism, interbreeding, and competition. In areas where the salmon are exotic, most concerns do not focus on interbreeding with other salmon species. Rather, they center on whether the escaped salmon will establish feral populations, reduce the reproductive success of native species through competition, alter the ecosystem in some unpredictable way, or transfer diseases (EAO, 1997).

However, a comprehensive evaluation of risks has concluded that the escape of Atlantic salmon pose very little or no risk to the environment of the Pacific Northwest, including through the mechanisms of colonization of salmonid habitat, competition with native species for forage, predation on indigenous species, and hybridization with other salmonids (Nash, 2001). Furthermore, another recent report finds little to no risk to "evolutionarily significant units" (ESUs) of Puget Sound chinook salmon and Hood Canal summer-run chum salmon arising from Atlantic salmon farms in Puget Sound (Waknitz et al., 2002). Authors of the latter study qualify their conclusion by stating that significant expansion of the industry may increase risks and some of the potential impacts might need to be reconsidered. Nevertheless, it should also be noted that Alaska Department of Fish and Game (ADFG) and others assert that Atlantic salmon may adversely effect native populations of Pacific salmon through mechanisms including colonization, habitat destruction, and competition (ADFG, 2002; Goldburg et al., 2001). ADFG recommends a gradual transition along the Pacific Coast to only land-based Atlantic salmon farming and storage operations. Research by Volpe and others (Volpe, 2000, 2001a, 2001b) suggests that Atlantic salmon may be capable of colonizing and persisting in coastal British Columbia river systems that are underutilized by native species.

In northeastern U.S., in contrast, aquaculture escapees were among the major threats to the Gulf of Maine distinct population segment (DPS) of Atlantic salmon identified by NOAA and USFWS ("Services") due to interactions between wild stocks and escapees. The Services noted that a large percentage of fish used at that time in aquaculture were of European origin and genetically different from native North American strains, and that North American strains used by the industry were genetically different from wild North American strains due to changes introduced through domestication. The Services further asserted that occurrences of adult escapees in Maine rivers were increasing commensurately with the growth of the aquaculture industry in Maine, and that government regulations and industry voluntary programs that existed at that time had not been effective in protecting wild stocks from aquaculture escapees. Considering scientific research including work suggesting that some level of introgression of European alleles may have already occurred, the Services concluded that "negative impacts to the DPS [from aquaculture escapes] can be reasonably anticipated to occur in Maine." The Services determined that the wild Gulf of Maine distinct population segment (DPS) of Atlantic salmon was in danger of extinction throughout its range and extended endangered status to this DPS (November 17, 2000; 65 FR 69459; available at http://www.nero.nmfs.gov/atsalmon/fr_fr.pdf).

Tilapia

The most commonly raised tilapia in the United States are blue (*Oreochromis aureus*), Nile (*O. niloticus*), Mozambique (*O. mossambicus*), and hybrids thereof. Native to Africa and the Middle East, tilapia have been introduced throughout the world as cultured species in temperate regions (Stickney, 2000). These freshwater fish of the family Cichlidae are primarily herbivores or omnivores. Feeding lower on the food chain has enhanced their popularity as a culture species (Stickney, 2000). Tilapia were first introduced to the Caribbean islands in the 1940s and then eventually were introduced to Latin America and the United States. In addition to production for foodfish, tilapia have been stocked in irrigation canals to control aquatic vegetation. Tilapia have also been used in the aquarium trade, as bait, as a sport fish, and as forage for warmwater predatory fish (Courtenay et al., 1984; Courtenay and Williams, 1992; Lee et al., 1980).

Tilapia have been found to be competitors with native species for spawning areas, food, and space (USGS, 2000a). Reports indicate that some streams, where blue tilapia are abundant, have lost most vegetation and nearly all native fish (USGS, 2000a). In Hawaii, tilapia is considered a threat to native species such as the striped mullet (*Mufil cephalus*; USGS, 2000b), and in California's Salton Sea area redbelly tilipia (*Tilapia zillii*) has been considered a significant factor in the decline of the desert pupfish (*Cyprinodon macularius*) (see Schoenherr, 1988).

Tilapia have also been introduced to other areas of the United States. Blue tilapia was evaluated for a number of beneficial uses by the Florida Game and Fresh Water Fish Commission. Although the Commission concluded that this species would be undesirable for stocking in Florida's public waters, the public removed fish from the study site, causing the tilapia to become established outside of the study site (Hale et al., 1995). Tilapia are now a commercially harvested species in Florida (Hale et al., 1995). During evaluation studies in North Carolina, blue and redbelly tilapia were inadvertently introduced into a reservoir. These species became established and led to the elimination of all aquatic macrophytes from the reservoir and declines in populations of several fish species (Crutchfield, 1995). In California, tilapia have become an important game fish, primarily in the Salton Sea, and their popularity with anglers is growing. Competition from and predation by Mozambique tilapia led to the extirpation of the High Rock Spring tui chub (*Gila bicolor*) from a California spring system. These tilapia were introduced from aquaculture facilities permitted by the California Department of Fish and Game (CDFG) in 1982. Inadequate screening of rearing facilities allowed tilapia to escape into the spring system (U.S. Department of Interior, Fish and Wildlife Service, 62 FR 49191-49193, September 19, 1997).

Because of its nonnative status, tilapia have been regulated by various States to prevent escapement and impacts on wild stocks of native species. Importation and movement of tilapia are regulated in the United States. According to Stickney (2000), the following states have some form of restriction on tilapia culture: Arizona, California, Colorado, Florida, Hawaii, Illinois, Louisiana, Missouri, Nevada, and Texas.

Several tilapia species and hybrids in the genus *Oreochromis* are raised at CAAP facilities in the scope of EPA's final regulation. EPA analysis suggests that the potential geographic distribution[4] of select tilapia species and hybrids may include California's San Joaquin Valley, southern California, southwestern Arizona, the Rio Grande River, and the Gulf Coast. Figures B-1 and B-2 (both in Appendix B) show, for all USGS 8-digit HUCs in the United States, the proportion of watershed area occupied by potential distribution, weighted by the number of distributional models (0-10 out of 10 models that had low underprediction errors) predicting presence in a grid cell. The potential geographic distribution of Mozambique, blue x Mozambique, and Wami River x Mozambique tilapia (Figures B-1a, B-1b, and B-1c) occurred in all these areas, in contrast to the more limited potential distributions of blue (Colorado River), Nile (Gulf Coast), and Wami River tilapia (southern Texas, Florida) (Figures B-2a, B-2b, and B-2c). Although these modeled distributions are considered robust, these should be regarded as a coarse view due to limited point-occurrence data. Furthermore, although it has been shown that convergent GARP predictions (locations where all models in the best-subset indicate potential presence) demonstrate high coincidence with areas of invasion/known occurrence, translating GARP output to a common numerical scale representing the likelihood of potential distribution, has not yet been done.

Data provided by facilities in the scope of EPA's final regulation indicate that several facilities raising one or more of the modeled species are located within the modeled potential distributions. As noted earlier, many States have established certain requirements relating to escapes of tilapia and/or non-native aquatic species in general; these States include some that fall within the modeled potential distribution area for tilapia. For example, most States in the area appear to require certain escape prevention measures. Mississippi State regulations, for instance, state that "[d]ue to the prolific nature of the Tilapia species, a fish barrier shall be designed to prevent the discharge of water containing Tilapia eggs, larvae, juveniles and adults from the permittee's property. Although Tilapia may not overwinter in Mississippi waters, precautions must be taken to limit their escape into native waters. This shall be accomplished by using a 1000 micron mesh screen"
http://www.mdac.state.ms.us/library/agencyinfo/regulations/administration/AquacultureActivities.pdf].
On the other hand, it appears that while several States have established reporting requirements for escaped non-native organisms, several States do not have such reporting requirements. However, facility-specific requirements regarding escape prevention, escape reporting, or other prevention or mitigation measures may be established through a NPDES permit Hochheimer and Meehan, (2004b). For further details of EPA's analysis and review of State requirements, see Kluza and McGuire (2004) and Hochheimer and Meehan, (2004b).

---

[4]The potential geographic distribution of a species in a region of interest may be estimated if the ecological niche of that species - defined based on nonrandom associations between point occurrence data for individuals of that species in its native range and ecological/environmental variables associated with the point occurrence data - as well as geographic information system coverages of the ecological/environmental variables for the region of interest, are available. EPA used the Genetic Algorithm for Rule-set Prediction (GARP) to model the potential geographic distribution of select tilapia species and hybrids. For further description of EPA's modeling analysis, see Kluza and McGuire (2004).

Other Issues Related to Escapes

As mentioned earlier, scientists have highlighted the need to carefully evaluate potential risks associated with the use of genetically modified (GM) organisms in aquatic animal production (e.g. Hedrick, 2001; Reichardt, 2000). Although the issue is being examined by commercial interests and under review by the Food and Drug Administration, there is no known current use of such organisms in U.S. aquaculture. Howard et al. (2004) studied mating competition and fitness between wild and genetically modified strains of Japanese medaka (*Oryzias latipes*); salmon growth hormones were added to a treatment group of male medaka to increase their size. The results showed GM males were more successful in mating with females, but produced offspring were less likely to survive than those sired by unaltered males. Howard et al. (2004) modeled these competing factors and the results suggest that if GM individuals are able to enter wild populations the transgene will spread, but will also ultimately lead to extinction of the population as offspring are less likely to survive[5].

7.3.3   Impacts from Drugs and Pesticides

7.3.3.1 Background

Drugs and pesticides are used at CAAP facilities as described in Section 7.2.5 for purposes including water quality maintenance, disinfection, anesthetization, and a variety of disease control and treatment purposes. Compounds reported in responses to EPA's detailed industry questionnaire to be used at CAAP facilities include AQUI-S, oxytetracycline, copper sulfate, formalin, hydrogen peroxide, and potassium permanganate and Chloramine-T.

Some drugs and pesticides used at CAAP facilities enter the environment with facility effluent following treatment. These compounds may affect non-target organisms in receiving environments, but any potential exposure depends on site-specific conditions and a number of general protections exist or have been instituted to mitigate potential impacts to non-target organisms. For example, approved drug and pesticide products are used only when needed for defined, specific purposes and for finite treatment durations. Furthermore, industry has developed a variety of quality assurance programs to promote a positive code of production practices that ensures a wholesome and safe product to consumers and the environment (Eirkson et al., 2000). In addition, FDA's environmental review processes result in drug label requirements, as necessary, that include directions on proper dilution before discharge and other conditions (e.g., filtration) that can control the amount of animal drug contained in effluents. FDA and EPA are also working on a formal agreement that would identify shared responsibilities for drug releases that pose an environmental risk.

FDA's Center for Veterinary Medicine (CVM), approves drugs for use in animals including aquatic animals under the Federal Food, Drug and Cosmetic Act (FFDCA). As part of the approval process, under the requirements of the National Environmental Policy Act (NEPA), CVM evaluates the environmental risks from the intended use of animal drugs and manages risks through labeling. FDA's authority applies to fish raised for human consumption, as well as to those fish used for stocking.

---

[5] Because the authors experiment with inserting genes of one species into another species, these organisms can be considered transgenic.

The FDA approval process may involve granting investigative new animal drug exemptions (INADs) from approved use for the purpose of establishing data on which to base approval of a drug. Through the investigative approval process, the sponsor agrees to conduct laboratory and field tests with the drug under the conditions and on the animals proposed for approval. These data are collected in the INAD and eventually submitted to a new animal drug application (NADA) to form the basis for CVM's approval or disapproval of the drug. Data collection for the drug approval includes data on the observed or anticipated environmental effects associated with the drug's use. In the case of drugs used on aquatic animals the most significant environmental effect anticipated with the drug's usage is the effect on the aquatic environment.

Because granting an INAD and approving a NADA are federal actions, the FDA must comply with NEPA as it carries out these processes. INADs and NADAs require submission of either a claim of categorical exclusion or an environmental assessment (EA). 21 C.F.R. 25.15, 21 C.F.R. 511.1(b)(10), 21 C.F.R. 514.1(b)(10). Most INADs are categorically excluded but require that investigators contact appropriate NPDES offices before discharging drugs in aquaculture wastewater. Most NADAs for aquaculture drugs require EAs. The EA facilitates the environmental component of FDA's "safety" review by providing information relevant to determining whether environmental consequences resulting from use of the new animal drug could adversely affect the health of humans or animals and possibly render the drug unsafe. An EA includes detailed information on the use of the drug, its environmental fate (e.g., water solubility, octanol/water partition coefficient, sediment/particulate absorption, degradation), toxicity (e.g., acute and chronic effects on daphnia, vegetation, and fish), exposure calculations, and risk characterization (Eirkson et al., 2000). FDA attempts to post all environmental assessments and supporting materials for environmental assessments for all FDA approved aquaculture drugs on the FDA/CVM web site (http://www.fda.gov/cvm/default.html).

FDA has made several guideline documents available to sponsors that detail protocols and procedures for environmental studies. These documents aid sponsors in developing the data and information needed to ensure environmental safety. Guidelines currently available to drug sponsors include FDA Guideline documents #61 (addressing FDA approval of new animal drugs for minor uses and for minor species) and #89 (addressing environmental impact assessments (EIAs) for veterinary medicinal products (VMPs). These documents are available on the FDA/CVM web site. In addition, FDA has announced the availability for public comment of an additional guideline document produced by the Veterinary International Cooperation on Harmonization (VICH) (69 FR 21152, April 21, 2004). Presently, this draft guideline addresses issues such as cumulative impacts and is available at http://vich.eudra.org/pdr/10_2003/gl38_st4.pdf. FDA anticipates that following a public comment process, this guideline, like FDA guideline documents #61 and #89, would also become available to sponsors.

Despite the existence of these general protections, evaluation of site-specific conditions to determine potential for environmental impact may be appropriate for several reasons. Current FDA environmental assessment protocols, and presumably environmental assessments upon which they were based, do not contemplate all possible discharge scenarios (e.g., cumulative effects from multiple dischargers and/or repeated applications or cumulative exposure to chemical stressors that share the same mechanism of action). Furthermore, potential impacts of drug/pesticide discharges on specific sensitive, threatened, or endangered species that may be present in receiving waters of particular facilities may not have been evaluated. The potential for adverse impacts on non-target wild organisms due to incidental poisoning (e.g., adverse impacts to scavengers from consumption of medicated prey or carcasses) may also not be addressed by existing environmental review processes. In addition, advances in scientific understanding of environmental fate, transport, and effects of certain compounds may not be reflected in all environmental assessments and label requirements. Also, only limited information on environmental effects may be available for drugs used under INAD exemptions, or under extra-label use provisions, and the need for site-specific consideration of potential impacts may exist. One example of where this was true was in the use of cypermethrin at a net pen facility in Maine. Through FDA's INAD program, cypermethrin was tested as a treatment for sea lice on cultured salmon. In the facility's 2000 draft NPDES permit, EPA allowed the facility to discharge cypermethrin into the surrounding waters. Through the information collected by FDA for the INAD program, EPA determined that cypermethrin use could potentially lead to adverse impacts to non-target organisms passing through or beyond the net pens' mixing zone, even at dosages lower than what is required for sea lice treatment. As a consequence, EPA found the use of cypermethrin to be inconsistent with Maine's water quality standards and did not authorize its use in the facility's final permit. FDA has concluded that further research is needed before cypermethrin can be approved for use at aquaculture facilities (USEPA, 2002d).

Reviews of drugs and pesticides used in aquaculture have been published (e.g., GESAMP, 1997; Boxall et al., 2001). Although these reviews may have a broader focus than on practices in the United States, certain observations may have relevance to the United States. GESAMP (1997) reviewed chemicals used in coastal aquaculture, which include chemicals associated with structural materials, soil and water treatments, antibacterial agents and other therapeutic drugs, pesticides, feed additives, and anaesthetics. According to this review, most aquaculture chemicals, if properly used, can be viewed as wholly beneficial with no adverse environmental impacts or increased risks to aquaculture workers. However, the authors identified several factors that could make the use of otherwise acceptable chemicals unsafe: these include excessive dosage and failure to provide for adequate neutralization or dilution prior to discharge. Among potential environmental issues of concern relating to improper use are chemical residues in wild fauna, toxic effects in non-target species, and antibacterial resistance. The authors conclude with recommended measures to promote safe and effective use of chemicals in coastal aquaculture.

7.3.3.2 Environmental Effects Literature

Various sources of information are available for assessing potential effects of aquaculture drugs and pesticides. In addition to scientific literature that may be published for any drug or pesticide used by CAAP facilities, FDA's CVM posts environmental assessments and supporting materials for environmental assessments for all FDA approved aquaculture drugs on the CVM web site at *http://www.fda.gov/cvm/default.html*.

The USGS Midwest Environmental Sciences Center, Drug Research and Development Program conducts research to support the approvals (Food and Drug Administration) or registrations (U.S. Environmental Protection Agency) of drugs intended for use in public fish husbandry and management. More information about this program is available at *http://www.umesc.usgs.gov/aquatic/drug_research.html*.

In connection with the CAAP rulemaking, EPA has informally compiled environmental fate and effects literature for each of a group of drugs and pesticides used at CAAP facilities, drawing from a wide range of sources, including those identified above. These compilations include information on trade names, generally reported use and dosage, and tabulations of toxicity test data from a variety of sources.

The informal EPA compilations are in the electronic docket accompanying EPA's final CAAP rule (http://cascade.epa.gov/RightSite/dk_public_home.htm).

Below are brief discussions of some environmental effects information available for several drugs and a pesticide that were commonly reported as being used at CAAP facilities surveyed for EPA's final CAAP rule. These discussion were drawn from the sources described above as well as other sources. Interested readers are urged to consult the sources of information identified above, the primary literature cited in this section, as well as any other current scientific literature that may be relevant to a reader's application.

Hydrogen Peroxide

Hydrogen peroxide (H_2O_2) is used under an INAD exemption to control bacterial gill disease (FDA, 1998), and has also been used as a "low regulatory priority" drug to control fungi on all species and life stages of fish, including eggs (JSA, 2000). Recommended treatment concentrations for fungus control are up to 500 ppm for up to 60 minutes (Syndel, 2003); treatment methodologies are still being developed (JSA, 2000).

The USGS has assessed the potential environmental fate and effects of hydrogen peroxide use for treating external fungal, bacterial, and parasitic diseases (Howe et al., 2000). According to this report, hydrogen peroxide concentrations used in aquaculture facilities range from approximately 50 – 1,000 ppm. Hatcheries generally dilute the hydrogen peroxide concentrations by 2 to 100,000-fold before discharge into surface water. The decomposition rate of hydrogen peroxide in natural waters ranges from a few minutes to longer than a week, depending on the chemical, biological, and physical factors of the aquatic ecosystem. In most cases, according to the report, hydrogen peroxide concentrations in receiving waters should reach background levels within a few hours after discharge from a hatchery. The report noted that dilute concentrations of hydrogen peroxide could have short-term impacts on a variety of aquatic plants and animals but concluded that no long-term effects such as altered species composition or population densities would occur due to brief exposure times. The report also noted that no persistent contaminants would be discharged into the environment or would accumulate in aquatic organisms as a result of hydrogen peroxide release into aquatic environments. Other studies have shown hydrogen peroxide to be toxic to a variety of non-target organisms when exposed for 96 hours at relatively low concentrations (Tetra Tech, 2003). Ninety-six hour toxicity tests on *Ceriodaphnia dubia* performed by the California Department of Fish and Game yielded a maximum allowable toxicant concentration (MATC) of 1.77 mg/L (CDFG, 2002). The MATC is defined as the maximum concentration at which a chemical can be present and not be toxic to the test organism. It is the range of concentrations between the lowest observed effect concentration (LOEC) and the no observed effect concentration (NOEC)[6].

---

[6] LOEC is the lowest treatment (i.e., test concentration) of a test substance that is statistically different in adverse effect on a specific population of test organisms from that observed in controls. NOEC is the highest treatment (i.e., test concentration) of a test substance that shows no statistical difference in adverse effect on a specific population of test organisms from that observed in controls. Note that the LOEC has to be less than the EC_{50}. If the LOEC is higher than the EC_{50}, then (1) the test has to be repeated to obtain a LOEC less than the EC_{50} or (2) the EC_{10} can be predicted from the dose-response curve (or the concentration-effect curve) (PBT Profiler, n.d.). EC_{50} (median effective concentration) is the statistically derived concentration of a substance in an environmental medium expected to produce a certain effect in 50 percent of test organisms in a given population under a defined set of conditions. The EC_{10} is the concentration where the effect is produced for 10 percent of the test organisms (McNaught and Wilkinson, 1997).

Formalin

Formalin is a solution of 37 percent formaldehyde gas by weight dissolved in water. The solution generally contains 10 to 15 percent methanol by weight to prevent polymerization (FDA, 1995). Formalin has been approved by FDA for use in several aquaculture applications under the trade names Formalin-F®, Paracide-F®, and Parasite-S®. Formalin is used to control fungi on finfish eggs and external parasites on finfish and shrimp. Treatment frequency, duration, and concentration varies with purpose of treatment, species, and culture conditions.

FDA has determined that no environmental impacts are expected, providing that treatment water is diluted adequately before being discharged to receiving waters (FDA, n.d.). FDA suggests that the concentrations of effluent from treatment tanks or raceways should be such that the concentration when diluted into the receiving waterbody is no greater than 1 ppm (FDA, 1995). In the finding of no significant impact for Parasite-S®, FDA requires a 10-fold dilution of finfish and penaeid shrimp treatment water and a 100-fold dilution of finfish egg treatment water, which should lead to a discharge concentration of no more than 25 ppm. FDA contended that additional in-stream dilution, infrequent use, and rapid degradation (formaldehyde, the active ingredient in formalin, is oxidized in the aquatic environment into formic acid and ultimately into carbon dioxide and water; the estimated half-life of formaldehyde in water is approximately 36 hours (FDA, 1995)) would render the discharged formalin below a level that causes significant environmental effects on aquatic animals (FDA, 1998). Directions for dilution of treatment water and additional environmental precautions are contained on the labeling of the product (FDA, n.d.).

In an environmental assessment performed in 1981 and submitted to FDA, U.S. Fish and Wildlife Service compiled results from several toxicity studies. USFWS noted that for most fish, formalin concentrations greater than 400-500 ppm cause mortality in 1 hour. No evidence of bioconcentration in fish tissue was found. Some fish prey organisms including daphnids (water fleas) and ostracods (seed shrimp) appear to be sensitive to formalin. In unusual circumstances, such as when effluent from fish treatment tanks or egg treatments are released into small, stagnant waterbodies, these releases would temporarily inhibit or damage phytoplankton and zooplankton populations, and contribute to hypoxic conditions. Any short-term inhibition or damage of these populations would be expected to recover rapidly (USFWS, 1981). Recent toxicity tests performed by the California Department of Fish and Game found the MATC is 2.7 ppm for the short term and 1.3 ppm for the long term (CDFG, 2002).

Oxytetracycline

Oxytetracycline has been approved by FDA to treat specific bacterial infections in catfish, salmonids, and lobster. It has also been approved to mark skeletal tissue in Pacific salmon so that resource management agencies can track salmon that are released to the wild. In the following listing of approved uses of oxytetracycline, minimum temperatures for treatment are specified (16.7°C for catfish and 9°C for salmonids) because temperatures below these minimums do not have approved withdrawal times. Clearance rates for oxytetracycline at lower temperatures and safe residual levels in tissues meant for human consumption are not known. Studies such as Meinertz et al. (2001) are being done to establish safe withdrawal times for treating aquatic animals which are meant for human consumption at lower temperatures. Other studies (e.g., Rigos et al., 2002) are being reported for determining the effectiveness and safety of treating other species with oxytetracycline.

Oxytetracycline is being used under an INAD for control of columnaris in walleye, vibriosis in summer flounder, and *Streptococcus* infection in tilapia (FDA, 1998). As stated earlier, the extralabel use of an FDA approved drug in or on feed is prohibited under AMDUCA. The Agency has granted regulatory discretion for the use of a medicated feed mixed according to the approval, for example oxytetracycline for salmon, to be used on or by the order of a veterinarian in an extra-label manner. The medicated feed cannot be modified in any way, for instance, it cannot be reformulated or repelleted. The medicated feed has to be labeled for the approved species and indication and only under a veterinarian's order can it be used extralabelly. For aquaculture this discretion applies only to those feeds approved for an aquaculture species.

In the *Finding of No Significant Impact for Terramycin (Oxytetracycline) Premix for Use in Lobster (NADA 38-439 C027)*, developed by Pfizer, Inc. (1987), it was determined that the potential for bioaccumulation or biomagnification of this compound in the environment was small (if it occurred at all). Pfizer (1987) also determined that there should be no development of resistance in environmental aquatic microorganisms resulting from the use of oxytetracycline at the levels prescribed under the NADA for use in lobster (Pfizer, Inc., 1987). This and other literature available from FDA and other sources suggest that environmental risk from therapeutic use of OTC for most applications is thought to be small and/or short term because OTC is likely to be well-chelated in the aquatic environment, among other reasons. It should be noted that a relatively small portion of oxytetracycline is actually retained by the target organisms. Instead, a large proportion of the drug administered in feed is thought to be lost to the environment (e.g., Smith et al.(1994); Smith (1996)). In addition, some researchers have further examined the possibility of the development of antimicrobial resistance in microorganisms (and other effects on microflora) in receiving water environments as a result of aquaculture medicated feed applications (see, e.g., Austin, 1985; Bebak-Williams et al., 2002). Please see these sources for further discussion of this issue.

Kerry et al. (1996) found detectable quantities of oxytetracycline beneath and near Atlantic salmon net pens and elevated levels of oxytetracycline-resistant bacteria. Capone et al. (1996) found oxytetracycline levels in sediments were correlated to facility usage. Capone observed oxytetracycline residues in edible wild crab meat collected under net cages that had undergone high levels of oxytetracycline treatment and noted that farm employees occasionally collected crabs for consumption. The levels observed in Capone's study exceeded FDA allowable tissue residue levels. Capone noted that health risks associated with ingesting food containing antibacterial residues are unclear and highly controversial but levels in excess of FDA levels suggest that the issue merits further attention. Although these and other studies show the presence of oxytetracycline in sediments or aquatic species below net pens, it is important to note that practices used at the time of the studies and the studies themselves are relatively old, that oxytetracycline use has declined since the studies were conducted, and that some of the high readings were from a facility that may have had anomalous application rates.

In sampling done at 13 hatcheries, antibiotics were only detected in effluent waters from five of the facilities (Thurman, et al., 2002). However, sampling was not timed to coincide with antibiotic treatments; antibiotic concentrations could be higher during periods of treatment. Oxytetracycline and sulfadimethoxine, the most frequently detected antibiotics, were found at concentrations in the range of 0.10- to 2.0 μg/L, with only two samples exceeding this range (10 μg/L oxytetracycline in one sample; >15 μg/L sulfadimethoxine in one sample). No antibiotics were found in samples taken from source water at the hatcheries. (Thurman, et al., 2002).

According to MacMillan (2003), no data currently exists to demonstrate a direct link between the use of antibiotics in aquaculture and the occurrence of human pathogens that are resistant to antibiotics. According to the author, only very limited data exists that documents the concentration of antibiotics in water as a consequence of the use of antibiotic medicated feed, and studies continue to be conducted to determine the potential impact of specific aquaculture drugs in the environment.

Copper

Copper, primarily in the form of copper sulfate ($CuSO_4$) and chelated copper (organically complexed copper) compounds, have been used for many years as a pesticide to control unwanted algae in ponds, tanks, and raceways. Copper compounds are also used as an antifoulant treatment for the nets used in net pen operations (Nash, 2001). Flexabar Aquatech's Flexgard is a latex algaecide dip designed for treating nets. The active ingredient in the dip is cuprous oxide (26%), which is highly toxic to fish and crustaceans (Flexabar Aquatech, n.d.; EAO, n.d.; PAN, n.d.).

Copper sulfate is also being tested (as an INAD) for use in the treatment of external parasites. More specifically, it is used to control bacterial diseases, fungal diseases, and external protozoan and metazoan parasites in finfish (Plumb, 1997). Copper sulfate has been used experimentally to treat fish parasites such as *Ichthyophthirius* (Ich), *Trichodina*, *Icthyobodo* (Costia), *Trichophyra*, *Chilodonella*, *Ambiphrya* (Scyphidia), *Apisoma* (Glossatella) and fungus (Masser and Jensen, 1991).

Copper is extremely toxic to aquatic organisms. It may be poisonous to trout and other fish, especially in soft or acidic waters, even when it is applied at recommended rates. Copper's toxicity to fish tends to decrease as water alkalinity increases. Fish eggs are more resistant to the toxic effects of copper than young fish fry. Copper is also toxic to aquatic invertebrate such as crabs, shrimp, and oysters (Extoxnet, 1996). For more information, refer to EPA's *Ambient Water Quality Criteria for Copper - 1984* (USEPA, 1985).

Copper is adsorbed to organic materials and to clay and mineral surfaces. The degree to which it is adsorbed depends on the acidity or alkalinity of the soil (Extoxnet, 1996). USDA cites Baudo et al. (1990) as saying: "The bioavailability of copper is regulated by water pH, sediment pH, sediment redox potential, acid volatile sulfides, sediment and waterborne organic carbon, particle size distribution, clay type and content, and cation exchange capacity of the sediment" (USDA, 1997).

Levels of copper around some net pen facilities may be elevated when it is used as an antifouling agent for the nets. According to Nash (2001), there is no evidence of long-term buildup of copper under salmon farms. As stated by Nash (2001), Lewis and Metaxas (1991) examined copper concentrations inside and immediately next to newly installed copper-treated nets at a net pen salmon farm in British Columbia. According to the authors, tidal exchange in and near net pens is important in maintaining low dissolved copper concentrations by preventing the accumulation of copper leached from nets. As reported in Nash (2001), Brooks (2000) stated that sediment copper concentrations at farms using copper treated nets were not always associated with the copper treatment itself but with other activities such as net washing, which can abrade copper-latex paint off the nets. Because of this, Brooks (2000) advised that any copper-treated nets should be washed and retreated at upland stations with any residual debris being buried at approved landfill sites.

Han et al. (2001) investigated the accumulation, distribution, and potential bioavailability of copper in sediments in catfish ponds that received weekly copper sulfate applications during summer growing seasons over 3 years. There was significant accumulation of copper (45.5 mg/kg/yr) in pond sediments at the end of the study, and the copper was not evenly distributed in pond sediments. Copper also accumulated with possible greater bioavailability in the copper sulfate treated ponds than non-treated ponds. Han et al. found that over time copper will redistribute through the soil as more and more stable fractions, thus reducing bioavailability.

Huggett et al. (2001) investigated the fate and effects of copper sulfate on non-target biota in streams that receive catfish pond effluent containing copper. Upstream and outfall samples did not adversely affect the test organisms used (*Hyalela azteca* and *Typha latifolia*), but the downstream samples did adversely affect *Hyalela azteca* survival. *Typha latifolia* germination and growth was not affected by the downstream sediment; however, shoot growth did decrease with increasing concentrations of copper. Effects of different sediment concentrations in this study may differ from other studies due to differences in sediment characteristics. Organic carbon and particle size, for example, greatly influence the bioavailability of copper in stream sediment.

7.3.4 Impacts from Pathogens

Although aquaculture facilities are not considered a source of human pathogens (see Section 7.2.6), it is possible that pathogens from other sources (e.g., mammals or birds) may be present in waste storage areas. MacMillan (2002) indicates that this is a unlikely source of risk. Nash (2003) also notes that there is little evidence to suggest that the accumulation of wastes from net-pen facilities is a source of human or environmental pathogens. Although some monitoring has showed a slight increase of fecal coliform near salmon farms, it is likely that these bacteria are from mammals or birds in the area.

It has also been suggested that aquaculture operations may be a source of disease to wild populations. Nash (2003) discusses the low risk that escaped Atlantic salmon would be vectors for the introduction of new, exotic pathogens into the Puget Sound area of Washington State. No new stocks of Atlantic salmon have been transferred into Washington since 1991, and any stocks transferred within the State must have a certification that they are disease-free, so it is not possible that Atlantic salmon already in the state would be vectors for exotic disease (Nash, 2003). Because all farmed salmon in Washington State are inspected annually for disease, they do not present a high risk for infection of wild stocks (Nash, 2003). While fish hatcheries may potentially be reservoirs of infectious agents (due to higher rearing densities and stress), little evidence suggests that disease transmission to wild stocks from hatcheries occurs routinely (Strom et al., n.d.).

In British Columbia, the Environmental Assessment Office (EAO) of British Columbia reported that between 1991 and 1995, 90 adult Atlantic salmon recovered in British Columbia and Alaska were examined to determine if they were infected with any diseases. Two fish were infected with *Aeromonas salmonicida*, the causative agent of furunculosis, and none of the fish contained unusual parasite infestations. Additionally, none of the tested fish were infected with common viral infections (Alverson and Ruggerone, 1998). In contrast, a recent study in British Columbia by Morton et al. (2004) showed an increased incidence of sea lice (*Lepeophtheirus salmonis*) in wild juvenile pink (*Oncorhynchus gorbuscha*) and chum (*O. keta*) salmon near net pen farms in the Broughton Archipelago of British Columbia. Morton et al. found that 90% of the pink and chum salmon sampled near net-pen farms were infected above the lethal limit for lice in the mobile stage. They also showed that the abundance of sea lice infestations were 8-times greater near net pens than in control sites, and that in areas with no farms the sea lice numbers were close to zero. According to the author, although the study does not provide a causal relationship between salmon farms, sea lice, and wild salmon infection rates, the findings do suggest the salmon farms are a source of sea lice in this region (Morton et al., 2004). It is important to remember that the density of net-pen aquaculture operations in the British Columbia area is much greater than that in the U.S. coastal waters of the Pacific Northwest.

7.4   REFERENCES

ADFG (Alaska Department of Fish and Game). 2002. *Atlantic Salmon.* White Paper prepared by the Alaska Department of Fish and Game. March 5, 2002.

Alverson, D.L., and G.T. Ruggerone. 1998. *Escaped Farm Salmon: Environmental and Ecological Concerns.* Environmental Assessment Office, Government of British Columbia. <http://www.eao.gov.bc.ca/PROJECT/AQUACULT/SALMON/ Report/final/vol3/vol3-b.htm>. Accessed March 2002.

AFS (American Fisheries Society). 1997. *Resource Policy Handbook: Introduction of Aquatic Species.* American Fisheries Society. <http://www.fisheries.org/resource/page13.htm>. Accessed January 2002.

Avault, J. 1996. *Fundamentals of Aquaculture.* AVA Publishing, Baton Rouge LA.

Austin, B. 1985. Antibiotic pollution from fish farms: effects on aquatic microflora. *Microbiological Sciences* 2(4):113-117.

Baudo, R., Giesey, J.P., and Muntau, H. 1990. *Sediments: Chemistry and Toxicity of In-Place Pollutants.* Lewis Publishers, Boston, MA. pp. 61-105.

Bebak-Williams, J., G. Bullock, and M.C. Carson. 2002. Oxytetracycline residues in a freshwater recirculating system. *Aquaculture* 205:221-230.

Blazer, V.S., and S.E. LaPatra. 2002. Pathogens of Cultured Fishes: Potential Risks to Wild Fish Populations. In *Aquaculture and the Environment in the United States,* ed. J. Tomasso. pp. 197-224. U.S. Aquaculture Society, A Chapter of the World Aquaculture Society, Baton Rouge, LA.

Boardman, G.D., V. Maillard, J. Nyland, G.J. Flick, and G.S. Libey. 1998. *The Characterization, Treatment, and Improvement of Aquacultural Effluents.* Departments of Civil and Environmental Engineering, Food Science and Technology, and Fisheries and Wildlife Sciences, Virginia Polytechnic Institute and State University, Blacksburg, VA.

Botts, W.F. 1999. *Aquatic Biological Investigation, Big Spring Creek.* Pennsylvania Department of Environmental Protection.

Botts, W.F. 2001. *Aquatic Biological Investigation, Yellow Breeches Creek.* Pennsylvania Department of Environmental Protection.

Boxall, A., L. Fogg, P. Blackwell, P. Kay, and E. Pemberton. 2001. *Review of Veterinary Medicines in the Environment.* R&D Technical Report, Environment Agency, Bristol, UK.

Breitburg, L. 1988. Effects of Turbidity on Prey Consumption by Striped Bass Larvae. *Transactions of the American Fisheries Society* 177:72-77.

Bron, J.E., C. Sommerville, R. Wootten, and G.H. Rae. 1993. Fallowing of marine Atlantic salmon, *Salmo salar* L., farms as a method for the control of sea lice, Lepeophtheirus salmonis (Kroyer, 1837). *Journal of Fish Diseases* 16:487-493.

Brooks, K.M. 2000. *Determination of copper loss rates from Flexgard XI™ treated nets in marine environments and evaluation of the resulting environmental risks.* Report to the Ministry of Environment for the BC Salmon Farmers Association, 1200 West Pender Street, Vancouver, BC V6E 2S9, 24 p. In: Nash, C.E., ed. 2001. *Technical Memorandum: The Net-Pen Salmon Farming Industry in the Pacific Northwest.* NMFS-NWFSC-49. U.S. Department of Commerce, National Oceanic and Atmospheric Administration, 125 p.

Brooks, K.M., and C.V.W. Mahnken. 2003a. Interactions of Atlantic salmon in the Pacific northwest environment II. Organic wastes. *Fisheries Research* 62:255-293.

Brooks, K.M., and C.V.W. Mahnken. 2003b. Interactions of Atlantic salmon in the Pacific northwest environment III. Accumulation of Zinc and Copper. *Fisheries Research* 62:295-305.

CDFG (California Department of Fish and Game). 2002. *Aquatic Toxicology Laboratory Report.* Lab Report to Brian Finlayson from The California Department of Fish and Game. Fax to Tetra Tech, Inc. on October 8, 2002.

Capone, D., D. Weston, V. Miller, and C. Shoemaker. 1996. Antibacterial residues in marine sediments and invertebrates following chemotherapy in aquaculture. *Aquaculture* 145:55-75.

Carlton, J.T. 2001. *Introduced Species in U.S. Coastal Waters. Environmental Impacts and Management Priorities.* Prepared for the Pew Oceans Commission, Arlington, VA., 28 pp.

Chen, S., S. Summerfelt, T. Losordo, and R. Malone. 2002. Recirculating Systems, Effluents, and Treatments. In *Aquaculture and the Environment in the United States*, ed. J. Tomasso, pp. 119-140. U.S. Aquaculture Society, A Chapter of the World Aquaculture Society, Baton Rouge, LA.

Cornell. 1998. *Treatment of Diseased Fish.* Cornell University, Cornell Veterinary Medicine, Ithaca. <http://web.vet.cornell.edu/public/FishDisease/resources/diagnostics/treatment.htm>. Accessed May 2001.

Courtenay, W.R., Jr., D.A. Hensley, J.N. Taylor, and J.A. McCann. 1984. Distribution of Exotic Fishes in the Continental United States. In *Distribution, Biology and Management of Exotic Fishes*, ed. W.R. Courtenay, Jr., and J.R. Stauffer, Jr., pp.41-77. Johns Hopkins University Press, Baltimore, MD.

Courtenay, W.R., Jr., and J.D. Williams. 1992. Dispersal of Exotic Species from Aquaculture Sources, with Emphasis on Freshwater Fishes. In *Dispersal of Living Organisms into Aquatic Ecosystems*, ed. A. Rosenfield, and R. Mann, pp. 49-81. Maryland Sea Grant Publication, College Park, MD.

Crutchfield, J.U. Jr., 1995. Establishment and expansion of redbelly Tilapia and blue Tilapia in a power plant cooling reservoir. In: *Uses and Effects of Cultured Fishes in Aquatic Ecosystems*, ed. H.L. Schramm, Jr., and R.G. Piper, pp. 452-461 American Fisheres Society Symposium 15, Proceedings of the International Symposium and Workshop on the Uses and Effects of Cultured Fishes in Aquatic Ecosystems, Albuquerque, New Mexico, 12-17 March 1994. American Fisheres Society, Bethesda, MD.

Dill, W.A., and A. J. Cordone. 1997. History and Status of Introduced Fishes in California, 1871-1996. Manuscript for Fish Bulletin of the California Department of Fish and Game. In United States Geological Survey, 2001. *Nonindigenous Fishes* – Ctenopharyngodon idella. <http://www.nas.er.usgs.gov/fishes/accounts/cyprinid/ct_idell.html>. Accessed March 2002.

EAO. n.d. *Salmon Aquaculture Review Final Report, Volume 3 Part D*. Environmental Assessment Office. <http://www.intrafish.com/laws-and-regulations/report_bc/vol3-d.htm>. Accessed May 2003.

EAO (Environmental Assessment Office). 1997. *Impacts of Farmed Salmon Escaping from Net Pens*. Environmental Assessment Office, Government of British Columbia. <http://www.eao.gov.bc.ca/project/aquacult/salmon/escape.htm>. Accessed March 2002.

Easton, M.D.L., D. Luszniak, and E. Von der Geest. 2002. Preliminary examination of contaminant loadings in farmed salmon, wild salmon and commercial salmon feed. *Chemosphere* 46(7):1053-1074.

Eirkson, C.E., R. Schnick, R. MacMillian, M.P. Gaikowski, and J. F. Hobson. 2000. *Aquaculture Effluents Containing Drugs and Chemicals*, second draft prepared July 23, 2000. Technical Subgroup for Drugs and Chemicals, Aquaculture Effluents Task Force, Joint Subcommittee on Aquaculture, Washington, DC.

Embeck, M.S. 2000. *Water Quality, Dissolved Oxygen and Sediment Investigation, PFBC Big Spring Culture Station, Big Spring Creek*. Pennsylvania Department of Environmental Protection.

Extoxnet. 1996. *Pesticide Information Profiles: Copper Sulfate*. Extension Toxicology Network, a cooperative effort of the University of California-Davis, Oregon State University, Michigan State University, Cornell University, and the University of Idaho. <http://www.ace.ace.orst.edu/info/extoxnet/pips/coppersu.htm>. Accessed May 2001.

FDA (Food and Drug Administration). n.d. *Parasite-S NADA 140-989: General Information*. <http://www.fda.gov/cvm/efoi/section2/140989.pdf>. Accessed May 2001.

FDA (Food and Drug Administration). 1995. *Environmental Impact Assessment for the Use of Formalin in the Control of External Parasites on Fish*. Food and Drug Administration. <http://www.fda.gov/cvm/efoi/ea/EA_Files/140-989EA.PDF>. Accessed April 2003.

FDA (Food and Drug Administration). 1997. NRSP-7 Holds Semi-Annual Committee Meeting. *FDA Veterinarian Newsletter* 12 (November/ December). <http://www.fda.gov/cvm/index/fdavet/1997/november.htm>. Accessed May 2001.

FDA (Food and Drug Administration). 1998. NRSP Holds Semi-Annual Committee Meeting. *FDA Veterinarian Newsletter* 13 (November/ December). <http://www.fda.gov/cvm/index/fdavet/1998/november.htm>. Accessed May 2001.

Findlay, R.H., L. Watling, and L.M. Mayer. 1995. Environmental impact of salmon net-pen culture on marine benthic communities in Maine: A case study. *Estuaries* 18:145-179.

Flexabar Aquatech. n.d. *Flexgard waterbase antifouling net coatings.* http://www.fishlink.com/flexgard/Flexgard_-_English/Application/application.html. Accessed May 2003.

Fries, L.T., and D.E. Bowles. 2002. Water quality and macroinvertebrate community structure associated with a sportfish hatchery outfall. *North American Journal of Aquaculture* 64:257-266.
Freeze, M., and S. Henderson. 1982. Distribution and status of the bighead carp and silver carp in Arkansas. *North American Journal of Fisheries Management* 2(2):197-200.

Fuller, P.L., L.G. Nico, and J.D. Williams. 1999. Nonindigenous Fishes Introduced into Inland Waters of the United States. *American Fisheries Society* Special publication 27. Bethesda, MD.

GESAMP. 1997. Towards safe and effective use of chemicals in coastal aquaculture. IMO/FAO/UNESCO-IOC/WMO/IAEA/UN/UNEP Joint Group of Experts on the Scientific Aspects of Marine Environmental Protection. *Reports and Studies GESAMP.* No. 65. London, IMO. 40 pp. <http://www.fao.org/docrep/meeting/003/w6435e.htm>. Accessed October 23, 2001.

Goldburg, R.J., M.S. Elliott, and R.L. Naylor. 2001. *Marine Aquaculture in the United States: Environmental Impacts and Policy Options.* Pew Oceans Commission, Arlington, VA., 33 pp.

Gregory, R.S., and T.G. Northcote. 1993. Surface, planktonic and benthic foraging by juvenile chinook salmon, *Oncorhynchus tshawytscha,* in turbid laboratory conditions. *Canadian Journal of Fisheries and Aquatic Sciences* 50:233-240.

Guerin, Martin, Mark E. Huntley, and Miguel Olaizola. 2003. Haematococcus astaxanthin: applications for human health and nutrition. *Trends in Biotechnology* 21(5):210-216.

Hale, M.M., J.E. Crumpton, and R.J. Schuler Jr., 1995. From sportfishing bust to commercial fishing boon: A history of the blue Tilapia in Florida. pp. 425-430 In: *Uses and Effects of Cultured Fishes in Aquatic Ecosystems.* Schramm, H.L., Jr., and R.G. Piper (editors). American Fisheres Society Symposium 15, Proceedings of the International Symposium and Workshop on the Uses and Effects of Cultured Fishes in Aquatic Ecosystems, Albuquerque, New Mexico, 12-17 March 1994. American Fisheres Society, Bethesda, MD.

Han, F.X., J.A. Hargreaves, W.L. Kingery, D.B. Huggett, and D.K. Schlenk. 2001. Accumulation, distribution, and toxicity of copper in sediments of catfish ponds receiving periodic copper sulfate applications. *Journal of Environmental Quality* 30:912-919.

Hedrick, P.W. 2001. Invasion of transgenes from salmon or other genetically modified organisms into natural populations. *Canadian Journal of Fisheries and Aquatic Science* 58:841-844.

Henry, J.G., and G.W. Heinke. 1996. *Environmental Science and Engineering.* 2d ed. pp. 327-328. Prentice-Hall, Inc., Upper Saddle River, NJ.

Hinshaw, J.M., and G. Fornshell. 2002. Effluents from Raceways. In *Aquaculture and the Environment in the United States*, ed. J. Tomasso, pp. 77-104. U.S. Aquaculture Society, A Chapter of the World Aquaculture Society, Baton Rouge, LA.

Hites, R.A., J.A. Foran, D.O. Carpenter, M.C. Hamilton, B.A. Knuth, and S.J. Schwager. 2004. Global assessment of organic contaminants in farmed salmon. *Science* 303:226-229

Hochheimer, J., A. Escobar, C. Moore, and C. Meehan. 2004. *Technical Memorandum: Metal and Other Pollutant Loadings Associated with TSS.* Tetra Tech, Inc., Fairfax, VA.

Hochheimer, J., and C. Meehan. 2004a. *Technical Memorandum: Summary of Analysis of Drug and Chemical Use at CAAP Facilities.* Tetra Tech, Inc., Fairfax, VA.

Hochheimer, J., and C. Meehan. 2004b. *Technical Memorandum: Summary of Information on NPDES Permits Relating to BMPs, Drug and Chemical Use, and Non-native Species.* Tetra Tech, Inc., Fairfax, VA.

Horak, D., 1995. Native and nonnative fish species used in State fisheries management programs in the United States. pp. 61-67 In: *Uses and Effects of Cultured Fishes in Aquatic Ecosystems.* ed., Schramm, H.L., Jr., and R.G. Piper. American Fisheries Society Symposium 15, Proceedings of the International Symposium and Workshop on the Uses and Effects of Cultured Fishes in Aquatic Ecosystems, Albuquerque, New Mexico, 12-17 March 1994. American Fisheries Society, Bethesda, MD.

Howard, R.D., J.A. DeWoody, and W.M. Muir. 2004. Transgenic male mating advantage provides opportunity for Trojan gene effect in fish. *Proceedings of the National Academy of Sciences* 101(9):2934-2938.

Howe, G.E., M.P. Gaikowski, L.J. Schmidt, J.J. Rach. 2000. *Environmental Assessment for the Proposed Use of Hydrogen Peroxide in Aquaculture for Treating External Fungal, Bacterial, and Parasitic Diseases of Cultured Fish.* U.S. Geological Society, LaCrosse, WI.

Huggett, D.B., D. Schlenk, and B.R. Griffin. 2001. Toxicity of copper in an oxic stream sediment receiving aquaculture effluent. *Chemosphere* 44:361-367.

IDEQ (Idaho Division of Environmental Quality). n.d. *Idaho Waste Management Guidelines for Aquaculture Operations.* Idaho Division of Environmental Quality. <http://www2.state.id.us/deq/ro_t/tro_water/aquacult_open.htm>. Accessed December 2001.

Jennings, D.P. 1988. Bighead carp (hypophthalmichthys nobilis): A biological synopsis. *Biological Report* 88(29). U.S. Fish and Wildlife Service.

JSA. 2000. *Draft: Aquaculture Effluents Containing Drugs and Chemicals*. Joint Subcommittee on Aquaculture, Technical Subgroup for Drugs and Chemicals, JSA, Aquaculture Effluents Task Force.

Kendra, W. 1991. Quality of salmonid hatchery effluents during a summer low-flow season. *Transactions of the American Fishery Society* 120:43-51.

Kerry, J., Coyne, R., Gilroy, D., Hiney, M., and Smith, P. 1996. Spatial distribution of oxytetracycline and elevated frequencies of oxytetracycline resistance in sediments beneath a marine salmon farm following oxytetracycline therapy. *Aquaculture* 145:31-39.

Kluza, D. and L. McGuire. 2004. *Revised Draft CAAP Non-native Species Analysis*. U.S. Environmental Protection Agency, Washington, DC.

Kohler, C.C., and W.R. Courtenay. n.d. *American Fisheries Society Position on Introductions of Aquatic Species*. American Fisheries Society, Introduced Fish Section. <http://www.afsifs.vt.edu/afspos.html>. Accessed January 2002.

Lackey, R.T. 1999. Salmon policy: Science, restoration, and reality. *Environmental Science and Policy* 2:369-379

La Rosa, T., S. Mirto, A. Mazzola, R.Danovaro. 2001. Differential responses of benthic microbes and meiofauna to fish-farm disturbance in coastal sediments. *Environmental Pollution* 112:427-434.

Lee, D.S., C.R. Gilbert, C.H. Hocutt, R.E. Jenkins, D.E. McAllister, and J.R. Stauffer, Jr. 1980. *Atlas of North American Freshwater Fishes*. North Carolina State Museum of Natural History, Raleigh, NC.

Lewis, A.G., and A. Metaxas. 1991. Concentrations of total dissolved copper in and near a copper-treated salmon net-pen. *Aquaculture* 99:269-276.

Loch, D.D., J. L. West, and D. G. Perlmutter. 1996. The effect of trout farm effluent on the taxa richness of benthic macroinvertebrates. *Aquaculture* 147:37-55.

MacMillan, J.R., R. Reimschuessel, B.A. Dixon, G.J. Flick, and E.S. Garrett. 2002. *Aquaculture Effluents and Human Pathogens: A Negligible Impact*. Contributed report by the Human Pathogens and Aquaculture Effluent Special Subgroup, submitted to the JSA Aquaculture Effluents Task Force, January 2002, 8 pp.

MacMillan, J.R. 2003. *Drugs Used in the U.S. Aquaculture Industry*. National Aquaculture Association, Charles Town, WV.

Masser, M.P., and J.W. Jensen. 1991. *Calculating Treatments for Ponds and Tanks*. SRAC Publication No. 410 Sourthern Regional Aquaculture Center, Stoneville, MS.

McGhie, T.K., C.M. Crawford, I.M. Mitchell, and D. O'Brien. 2000. The degradation of fish-cage waste in sediments during fallowing. *Aquaculture* 187:351-366.

McGuire, L. 2003. *Memorandum: Conference call with William Botts and Mark Embeck, Water Pollution Biologists*. Water Management Program, South Central Region, Pennsylvania Department of Environmental Protection (PA DEP), 2:30-4:00 pm June 25, 2003.

McGuire, L. 2004. *Draft Memorandum: Discussions on PCBs*, January 12, 2004.

McNaught, A.D., and A. Wilkinson. 1997. *Compendium of Chemical Terminology: IUPAC Recommendations*. <http://www.chemsoc.org/cgi-shell/empower.exe?DB=goldbook>. Accessed May 2004.

Meinertz, J.R., M.P. Gaikowski, G.R. Stehly, W.H. Gingerich, and J.A. Evered. 2001. Oxytetracycline depletion from skin-on fillet tissue of coho salmon fed oxytetracycline medicated feed in freshwater at temperatures less than 9 °C. *Aquaculture* 198:29-39.

Moring, J.R., 1989. Documentation of unaccounted-for losses of chinook salmon from saltwater cages. *The Progressive Fish-Culturist* 51(3):173-176.

Morton, A., R. Routledge, C. Peet, and A. Ladwig. 2004. Sea lice (*Lepeophtheirus salmonis*) infection rates on juvenile pink (*Oncorhynchus gorbuscha*) and chum (*Oncorhynchus keta*) salmon in the nearshore marine environment of British Columbia, Canada. *Canadian Journal of Fisheries and Aquatic Sciences* 61:147-157.

Mosso, D., J. Jarcum, and J. Hochheimer. 2003. *Water Quality and Sediment/Benthic Impacts and Modeling Tools Used in Assessment at Net Pen Facilities*. Tetra Tech, Inc., Fairfax, Virginia.

Nash, C.E., ed. 2001. *The net-pen salmon farming industry in the Pacific Northwest*. U.S. Department of Commerce, NOAA Technical Memorandum NMFS-NWFSC-49. 125 pp.

Nash, C.E. 2003. Interactions of Atlantic salmon in the Pacific Northwest VI. A synopsis of the risk and uncertainty. *Fisheries Research* 62:339-347.

Naylor, R.L., S.L. Williams, and D.R. Strong. 2001. Aquaculture – A Gateway for Exotic Species. *Science* (294):1655-1656.

NOAA (National Oceanic and Atmospheric Administration). 1999. *The Environmental Impacts of Aquaculture*. A White Paper prepared by NOAA Marine Sanctuaries Division, National Ocean Service and Office of Habitat Conservation, Northeast Region, National Marine Fisheries Service, July 1999, 29 pp.

PAN (Pesticides Action Network North America). n.d. *Acute Toxicity Studies for Copper*. <http://www.pesticideinfo.org/PCW/List_AquireAcuteSum.jsp?CAS_No=7440-50-8&Rec_Id=PC33553>. Accessed May 2003.

PBT Profiler. nd. *Definitions*. Developed by the Environmental Science Center under contract to the Office of Pollution Prevention and Toxics, U.S. Environmental Protection Agency. <http://www.pbtprofiler.net/Details.asp>. Accessed May 2004.

Peterson, L.K., J.M. D'Auria, B.A. McKeown, K. Moore, and M. Shum. 1991. Copper levels in the muscle and liver tissue of farmed chinook salmon, *Oncorhynchus tshawytscha*. *Aquaculture* 99:105–115.

Pfizer, Inc. 1987. *Finding of No Significant Impact for Terramycin (Oxytetracycline) Premix for Use in Lobster (NADA 38-439 C027)*. Pfizer, Inc., New York, NY.

Pflieger, W. L. 1975. The Fishes of Missouri. In *United States Geological Survey, 2001*. Nonindigenous Fishes – *Ctenopharyngodon idella*, 343 pp. Missouri Department of Conservation, Jefferson City, MO. <http://www.nas.er.usgs.gov/fishes/accounts/cyprinid/ct_idell.html>. Accessed March 2002.

RAC (Regional Aquaculture Centers). 1998. *Compendium Report: 1989-1996*. Regional Aquaculture Centers.

Redding, J.M., C.B. Schreck, and F.H. Everest. 1987. Physiological effects on coho salmon and steelhead of exposure to suspended solids. *Transactions American Fisheries Society* 116:737-744.

Reichhardt, T. 2000. Will souped up salmon sink or swim. *Nature* 406:10-12.

Rigos, G., M. Alexis, A. Andriopoulou, and I. Nengas. 2002. Pharmacokinetics and tissue distribution of oxytetracycline in sea bass, *Dicentrarchus labrax*, at two water temperatures. *Aquaculture* 210:59-67.

Schoenherr, A. 1988. A review of the life history of the desert pupfish, *Cyprinodon macularius*. *Bulletin of the Southern California Academy of Science* 87:104-134.

Schomack, D., and H. Gray, 2002. Letter from Hon. Dennis Schomack, Chair, U.S. Section, International Joint Commission, and The Rt. Hon. Herb Gray, PC, QC, Chair, Canadian Section, International Joint Commission, to Honorable Colin Powell, Secretary of State, and The Honorable Bill Graham, Minister of Foreign Affairs. Letter dated July 5, 2002.

Schueler, T.R., and H.K. Holland. 2000. *The Practice of Watershed Protection*. Center for Watershed Protection, Ellicott City, MD.

SCAN (Scientific Committee on Animal Nutrition). 2002a. *Report of the Scientific Committee for Animal Nutrition on the Use of Astaxanthin-rich Phaffia Rhodozyma in Feedingstuffs for Salmon and Trout*. European Commission, Health and Consumer Protection Directorate-General.

SCAN (Scientific Committee on Animal Nutrition). 2002b. *Opinion of the Scientific Committee on Animal Nutrition on the Use of Canthaxanthin in Feedingstuffs for Salmon and Trout, Laying Hens, and Other Poultry*. European Commission, Health and Consumer Protection Directorate-General.

Smith, P. 1996. Is sediment deposition the dominant fate of oxytetracycline used in marine salmonid farms: a review of available evidence. *Aquaculture* 146:157-169.

Smith, P., J. Donlon, R. Coyne, and D.J. Cazabon. 1994. Fate of oxytetracycline in a fresh water fish farm: influence of effluent treatment systems. *Aquaculture* 120:319-325.

Snowdon, M. 2003. *Feed analysis values: Explanation of terms.* New Brunswick Department of Agriculture, Fisheries and Aquaculture. New Brunswick, Canada.

Stickney, R.R. 2000. Tilapia Culture. In *Encyclopedia of Aquaculture*, ed., R.R. Stickney, pp. 934-941. John Wiley and Sons, Inc., NY.

Strain, P.M., D.J. Wildish, and P.A. Yeats. 1995. The application of simple models of nutrient loading and oxygen demand to the management of a marine tidal inlet. *Marine Pollution Bulletin* 30:253-261.

Strom, M.S., L.D. Rhodes, and L.W. Harrell. n.d. *A Review of the Interactions between Hatchery and Wild Salmonids and Possible Spread of Infectious Disease.* Fish Health/Microbiology Team, Integrative Fish Biology Program, Resource Enhancement and Utilization Technologies Division, Northwest Fisheries Science Center.

Syndel. 2003. *Perox-Aid.* Syndel International, Inc <http://www.syndel.com/d_p_f_s/perox-aid_info_sheet.html>. Accessed July 2003.

Taylor, J.N., W.R. Courtenay, Jr., and J.A. McCann. 1984. Known impact of exotic fishes in the continental United States. In *Distribution, Biology, and Management of Exotic Fish*, ed. W.R. Courtenay, Jr., and J.R. Stauffer, pp. 322-373. Johns Hopkins Press, Baltimore, MD.

Tetra Tech, Inc. 2001a. *Sampling Episode Report Clear Springs Foods, Inc. Box Canyon Facility, Episode 6297.* Tetra Tech, Inc., Fairfax, VA.

Tetra Tech, Inc. 2001b. *Sampling Episode Report, Fins Technology, Turners Falls, Massachusetts, Episode 6439, April 23-28, 2001.* Tetra Tech, Inc., Fairfax, VA.

Tetra Tech, Inc. 2002a. *Sampling Episode Report, Harrietta Hatchery, Harrietta, MI, Episode 6460.* Tetra Tech, Inc., Fairfax, VA.

Tetra Tech, Inc. 2002b. *Site Visit Report for Acadia Aquaculture (ME).* Tetra Tech, Inc., Fairfax, VA.

Tetra Tech, Inc. 2002c. *Site Visit Report for Harlingen Shrimp Farm, Arroyo Shrimp Farm, Loma Alta Shrimp Farm (TX).* Tetra Tech, Inc., Fairfax, VA.

Tetra Tech, Inc. 2002d. *Site Visit Report for Heritage Salmon (ME).* Tetra Tech, Inc., Fairfax, VA.

Tetra Tech, Inc. 2003. *Technical Memorandum: Summary of Total Amount of Drugs/Chemicals Used in 2001 as Reported by Facilities in the Detailed Survey.* Tetra Tech, Inc., Fairfax VA.

Thurman, E.M, J.E. Dietze, and E.A. Scribner. 2002. *Occurrence of Antibiotics in Water from Fish Hatcheries.* U.S. Geological Society, Toxic Substances Hydrology Program.

USDA. 1997. *Environmental assessment for proposed approval of copper sulfate for use in aquaculture for control of waterborne parasitic, bacterial, and fungal diseases of cultured food fish.* Stuttgart National Aquacultural Research Center, United States Department of Agriculture, Stuttgart, AR.

USEPA (U.S. Environmental Protection Agency). 1985. *Ambient Water Quality Criteria for Copper — 1984.* EPA 440-5-84-031. U.S. Environmental Protection Agency, Washington, DC.

USEPA (U.S. Environmental Protection Agency). 1986. *Ambient Water Quality Criteria for Dissolved Oxygen.* EPA 440-5-86-003. U.S. Environmental Protection Agency, Washington, DC.

USEPA (U.S. Environmental Protection Agency). 2000. *National Water Quality Inventory: 1998 Report to Congress.* EPA 841-R-00-001. U.S. Environmental Protection Agency, Office of Water, Washington, DC. <http://www.epa.gov/305b/98report/toc.html>. Accessed December 2001.

USEPA (U.S. Environmental Protection Agency). 2002a. *Detailed Questionnaire for the Aquatic Animal Production Industry.* OMB Control No. 2040-0240. U.S. Environmental Protection Agency, Washington, DC.

USEPA (U.S. Environmental Protection Agency). 2002b. *Economic and Environmental Impact Analysis of Proposed Effluent Limitations Guidelines and Standards for the Concentrated Aquatic Animal Production Industry Point Source Category.* EPA 821-R-02-015. U.S. Environmental Protection Agency, Washington, DC.

USEPA (U.S. Environmental Protection Agency). 2002c. *Ecological Risk Assessment for the Middle Snake River, Idaho.* National Center for Environmental Assessment, Washington, DC; EPA-600-R-01-017.

USEPA (U.S. Environmental Protection Agency). 2002d. *Response to Comments in Regard to Authorization to Discharge Under the National Pollutant Discharge Elimination System.* Prepared by EPA-Region 1, Boston, MA. 64 pp.

USEPA (U.S. Environmental Protection Agency). 2004. *Development Document for the Final Effluent Limitations Guidelines and Standards for the Concentrated Aquatic Animal Production Point Source Category.* EPA 821-R-04-012. U.S. Environmental Protection Agency, Washington, DC.

USFWS (U.S. Fish and Wildlife Service). 1981. *Environmental Assessment: Use of Formalin in Fish Culture as a Parasiticide and Fungicide.* Submitted to Master File 3543. U.S. Fish and Wildlife Service.

USGS (U.S. Geological Survey). n.d. *Chemical contamination of hatchery fish feed.* <http://wfrc.usgs.gov/research/contaminants/STSeelye4.htm>. Accessed June 2002.

USGS (U.S. Geological Survey). 2000a. *Nonindigenous Fishes* — Oreochromis aureus. United States Geological Survey, Nonindigenous Aquatic Species. <http://www.nas.er.usgs.gov/fishes/accounts/cichlida/or_aureu.html>. Accessed March 2002.

USGS (U.S. Geological Survey). 2000b. *Nonindigenous Fishes* — Oreochromis mossambicus. United States Geological Survey, Nonindigenous Aquatic Species. <http://www.nas.er.usgs.gov/fishes/accounts/cichlida/or_mossa.html>. Accessed March 2002.

Utah DEQ. 2000. *Mantua Reservoir TMDL*. Utah Department of Environmental Quality.

VDEQ (Virginia Department of Environmental Quality & Virginia Department of Conservation and Recreation). 2002. *Benthic TMDL Reports for Six Impaired Segments in the Potomac-Shenandoah and James River Basins*. Available online at http://www.deq.state.va.us/tmdl/apptmdls/shenrvr/trout.pdf.

Volpe, J.P., E.B. Taylor, D.W. Rimmer, and B.W. Glickman. 2000. Evidence of natural reproduction of aquaculture-escaped Atlantic salmon in a coastal British Columbia river. *Conservation Biology* 14(June):899-903.

Volpe, J.P., B.R. Anholt, and B.W. Gilckman, 2001a. Competition among juvenile Atlantic salmon (*Salmo salar*) and steelhead (*Oncorhynchus mykiss*): Relevance to invasion potential in British Columbia. *Canadian Journal of Fisheries and Aquatic Science* 58:197-207.

Volpe, J.P., B.W. Gilckman, and B.R. Anholt, 2001b. Reproduction of aquaculture Atlantic salmon in a controlled stream channel on Vancouver Island, British Columbia. *Transactions of the American Fisheries Society* 130:489-494.

Waknitz, F.W., T.J. Tynan, C.E. Nash, R.N. Iwamoto, and L.G. Rutter. 2002. Review of Potential Impacts of Atlantic Salmon Culture on Puget Sound Chinook Salmon and Hood Canal Summer-Run Chum Salmon Evolutionarily Significant Units. NMFS-NWFSC-52. U.S. Department of Commerce, National Oceanic and Atmospheric Administration. 83p.

Wetzel, R.G. 1983. *Limnology.* 2d ed. Saunders College Publishing, Philadelphia, PA. 767 pp. and appendices.

Whelan, G.E. 1999. Managing Effluents from an Intensive Fish Culture Facility: The Platte River State Fish Hatchery Case History. Appendix 7 in: *Addressing Concerns for Water Quality Impacts from Large-Scale Great Lakes Aquaculture*. Based on a Roundtable co-hosted by the Habitat Advisory Board of the Great Lakes Fishery Commission and the Great Lakes Water Quality Board of the International Joint Commission. Available online at: <http://www.ijc.org/php/publications/html/aquaculture/>. Accessed March 2004.

CHAPTER 8

ENVIRONMENTAL BENEFITS OF FINAL REGULATION

8.1   INTRODUCTION

The final effluent limitations guidelines for concentrated aquatic animal production (CAAP) facilities requires permittees subject to the final rule to establish practices to control solids (e.g., employ efficient feed management and feeding strategies to minimize potential discharges of uneaten feed and waste products to waters of the U.S.), properly store drugs, pesticides, and feed; regularly maintain, routinely inspect and promptly repair any damage to the production system and wastewater treatment system; maintain certain records and provide for employee training. In addition, the final regulation establishes certain reporting requirements relating to the use of investigational new animal drugs (INADs) or approved drugs used in an extralabel fashion and relating to failure in or damage to the structure of an aquatic animal containment system. Please see the final regulatory text, as well as Chapter 4 of this document, for specifics on final regulatory requirements. These requirements, according to EPA loadings estimates, will reduce facility discharges of TSS and pollutants associated with the reduction in TSS discharges including total nitrogen (TN), total phosphorus (TP), and biochemical oxygen demand (BOD). EPA has also found that reductions in TSS will lead to reductions for feed contaminants (e.g., metals) as a result of these final requirements. Pollutant load estimates are discussed in Chapter 10 of USEPA (2004).

Reductions in these loadings (TSS, TN, TP, BOD, metals, and feed contaminants) could affect water quality, the uses supported by varying levels of water quality, and other aquatic environmental variables (e.g., primary production and populations or assemblages of native organisms in the receiving waters of regulated facilities). These impacts may result in environmental benefits, some of which have quantifiable, monetizable value to society. For the final regulation, EPA has only monetized benefits from recreational and non-use benefits associated with water quality improvements from reductions in TSS, TN, TP, and BOD (Table 8-1 provides a summary of the environmental benefits of the final regulation). EPA did not attempt to estimate benefits from possible reductions of feed contaminants discharged to receiving waters that may arise from reporting requirements. EPA anticipates that other requirements of the final rule will benefit the environment. For example, EPA believes that the requirement to notify the permitting authority of the use of INADs and approved drugs used in an extralabel fashion is necessary to ensure that any potential risk to the environment resulting from the use of these drugs can be addressed with site-specific remedies where authorized. This provides the permitting authority with the opportunity to monitor or control the discharge of the drugs while the drugs are being applied. EPA also anticipates that requirements relating to structural integrity of production and wastewater treatment system will also result in reduced losses of material to waters of the U.S. However, EPA has not attempted to monetize these anticipated benefits. This chapter will present a summary of results and the methods EPA used to evaluate only the potential monetized environmental benefits of the regulation.

Table 8-1
Summary of Environmental Benefits of the Final Rule

Type of Benefit	Environmental Benefits (Thousands of $2003)
Recreational and non-use benefits from improved water quality resulting from final requirements to establish solids control and feed management practices	Option A: $84 Option B: $94 - $118 FINAL OPTION: $66 - $99
Reduced discharge of feed additives/contaminants (e.g, metals, PCBs, other trace substances) from final requirements to establish solids control and feed management practices	not monetized
Better opportunity for permitting authority to evaluate potential for environmental risk and establish site-specific remedies, as appropriate, from required reporting to permitting authority of certain drug uses as described in rule	not monetized
Improved containment of materials and response to containment system damage or failure resulting from final reporting requirements regarding damage/failure of containment system and final requirements regarding maintenance practices	not monetized
Requirements for practices regarding proper materials (drugs, pesticides, and feed) storage and management of spilled materials	not monetized

8.2 MONETIZED BENEFITS

8.2.1 Overview of Method

EPA monetized water quality benefits resulting from the final rule using a combination of engineering, scientific, and economic analyses. EPA used engineering analyses to estimate reductions in TSS, nitrogen, phosphorus, and BOD loads from affected facilities under the final regulation (Table 10.6-2 and Chapter 10 of the TDD), water quality modeling tools to simulate the effects of these loading reductions on the water quality of receiving waters to which regulated facilities discharge, and economic valuation tools to estimate the monetary value that society places on these changes in water quality. Instead of assessing water quality impacts for each individual facility in the regulated population, EPA developed a set of water quality modeling "case studies" that were used to represent groups of facilities in the regulated population. EPA assumed that facilities in each group would experience water quality responses to the final regulation similar to those of the representative case study, and used these estimated water quality responses as the basis for estimating monetized benefits at regulated facilities.

EPA used facility-specific data (from the detailed industry questionnaire, see following section) for use in estimating and monetizing national environmental benefits[7]. The detailed questionnaire data provided specific information about each facility, such as facility configuration, feeding rates, system flows, and facility location. Some of this information was used in the engineering analysis to provide pollutant loading estimates under baseline and different regulatory option scenarios. These pollutant load estimates were used with the flow data in the QUAL2E modeling. Facility location information was used to determine the relationship of the facility effluent to the receiving water and to evaluate geographic relationships among facilities.

The following sections describe data sources, procedure for estimating national benefits based on results for a subset of facilities, and EPA's application of the water quality and economic valuation tools to estimate national environmental benefits from the final regulatory requirements for TSS.

8.2.2 Detailed Questionnaire Data

As described in the Notice of Data Availability for the CAAP rule (68 FR 75072, December 29, 2003), EPA developed a detailed questionnaire to collect data from CAAP facilities as the basis for estimating costs and benefits of the final CAAP rule. The detailed questionnaire itself, EPA's intended use of the data, sampling design, summary of responses received, and other aspects of the detailed questionnaire are all available in the administrative record for the final CAAP rule (USEPA, 2002; USEPA, 2004). Briefly, EPA mailed detailed questionnaires to a stratified random sample of aquaculture facilities. Of these, a large proportion of questionnaires were completed, returned to EPA, and were able to be used in subsequent analyses. Because EPA selected these facilities using a statistical design (see Appendix A of the Technical Development Document for the proposed rule), the responses allowed EPA to build a database to use for estimating population characteristics. That is, EPA had classified aquaculture facilities into strata defined by facility type (commercial, government, research, or tribal), the predominant species, and predominant production, and a sample was drawn from the population of aquaculture facilities ensuring sufficient representation of facilities in each of the strata. For national (i.e., population) estimates, EPA applied survey weights to the facility responses that incorporate the statistical probability of a particular facility being selected to receive the detailed questionnaire and adjust for non-responses. In this case, a survey weight of "3" means that the facility represents itself and two others in the population. As with cost and loading analyses for the final rule, EPA uses the detailed questionnaire database and sample weights as the basis for analysis of the environmental benefits of the regulation. In the subsequent discussions in this Chapter, a "detailed questionnaire facility" refers to a facility which completed and returned a detailed questionnaire which was able to be used in EPA's analyses.

---

[7]This approach extends the approach used in the environmental benefits analysis for the proposed rule by configuring water quality models to better represent the varying characteristics of CAAP facilities in the scope of the final regulation. For the proposed rule, EPA estimated water quality-related benefits for flow-through and recirculating facilities by simulating the water quality impacts of varying "model" facility discharge scenarios on a single "prototype" stream reach. EPA used the results of these scenarios to estimate national environmental benefits (see the proposal EA for additional details). EPA's current modeling approach used estimates for facility-specific effluent concentrations and flow rates to more accurately represent the contribution of individual facilities to receiving water changes. The revised approach also used receiving stream characteristics that represent background water quality conditions and hydraulic properties of receiving waters to which CAAP facilities discharge.

8.2.3 Extrapolation Framework

EPA developed a method to guide selection and development of a limited number of water quality modeling case studies that would be representative of the facilities for which EPA received a usable detailed questionnaire and that were in the scope of the final regulation. First, EPA assumed that water quality improvements at regulated facilities will be driven by three factors: the relative change in pollutant loadings resulting from the regulation, the concentration of the pollutants in the discharge, and the amount of dilution that occurs when the discharge enters the receiving water[8]. EPA then assigned each detailed questionnaire facility in the scope of the final regulation with non-zero load reductions into one of eight possible subgroups ("extrapolation categories") based on their value ("Low" or "High") for each of these three factors. EPA determined each facility's value for each factor as follows:

8.2.3.1 Factor 1: Regulatory Changes in Pollutant Loadings

EPA assumed that the percentage of pollutant load reductions at a given facility under the final rule would be one important factor in determining the magnitude of water quality response to the regulation. Water bodies receiving discharge from facilities that experience a large percentage reduction in pollutant loads as a result of the final regulation have the potential to experience larger water quality responses than those receiving discharge from facilities that experience smaller load reductions. EPA estimated percent TSS load reductions for each facility based on data provided by facilities in the detailed questionnaires, and with supplemental engineering analysis of the facility-provided data, where necessary. Percent load reductions for in-scope, detailed questionnaire facilities ranged from less than 1% to greater than 50%. A full description of load estimate calculations is provided in Chapter 10 of the Technical Development Document. The median percent TSS load reductions value for the detailed questionnaire facilities[9] was used as the threshold between the "Low" and "High" categories for this factor.

8.2.3.2 Factor 2: Pollutant Concentration in Discharge

EPA assumed that the baseline TSS concentrations of pollutants in facility effluents discharged to receiving waters would be a second important factor in determining the magnitude of water quality response to the regulation. EPA assumed that if baseline TSS concentrations of pollutants were low, then water quality responses to reductions in pollutant concentrations would be limited; conversely, if baseline TSS concentrations of pollutants were high, then water quality responses to reductions had the potential to be larger. EPA assessed baseline TSS concentrations in flow-through and recirculating system

---

[8]EPA informally evaluated the relationship between the three factors - % TSS load reduction, baseline TSS concentration, and dilution ratio - and water quality response by analyzing the relationship between these three factors and an output from a set of water quality modeling simulations. EPA performed multiple regression analyses between the three explanatory factors and an aggregate measure of water quality response (change in WQI6, a metric described later in this document) using four different model specifications. Using a linear model specification, the three explanatory factors explained 89% of the variation; using a log-log specification, the three explanatory factors explained 99% of the variation in d(WQI6). See McGuire (2004c).

[9]Median value estimate for percent TSS load reduction from median value indicated on April 2, 2004 Tetra Tech spreadsheet (Tetra Tech, 2004).

facilities that provided detailed production, feeding, facility configuration, and flow rate data. Baseline effluent TSS concentrations for in-scope, detailed questionnaire facilities ranged from less than 1 mg/L to greater than 40 mg/L. A more detailed explanation is available in Chapter 10 of the Technical Development Document. Again, EPA used the median baseline TSS concentration as the threshold between the "Low" and "High" extrapolation categories[10].

8.2.3.3 Factor 3: Dilution of Discharge in Receiving Water

EPA assumed that the amount of dilution that occurs in the receiving waters to which a facility discharges would be a third important factor determining the magnitude of water quality response to the regulation. Dilution ratios were estimated by dividing the facility flow by the sum of the receiving water flow and the facility flow. If the effluent flow rate is small relative to receiving water flow (low dilution ratio), then water quality response to the regulation is likely to be smaller than if the effluent flow rate is large relative to receiving water flow rate. EPA obtained effluent flow rates from data provided by facilities in the detailed questionnaires. EPA obtained receiving water flow data from a database of estimated mean annual and summer flows for all streams in the "Reach File 3" national stream reach network (USEPA, 2003; McGuire, 2004b). Due to limitations in the quality of geographic referencing data available (e.g., latitude and longitude coordinates whose accuracy could not be established) and other data limitations, EPA was able to estimate receiving water flow rates and dilution ratios at a subset of detailed questionnaire facilities. Dilution factors ranged from less than 0.01 to 0.90 for in-scope, detailed questionnaire facilities for which dilution ratios could be determined. Again, EPA used the median value as the threshold between the "Low" and "High" extrapolation categories for this factor[11].

Eight distinct "extrapolation categories" can be generated based on different combinations of the above three factor values. For example, a category defined by "Low" percent TSS load reduction, "Low" baseline TSS concentration, and "Low" dilution ratio is designated "LLL;" similarly, the categories "LLH", "LHL," "LHH," "HLL," "HLH," "HHL," "HHH" can be defined (Table 8-2). Using the thresholds described above and the detailed questionnaire data, EPA assigned each in-scope, detailed questionnaire facility with non-zero load reductions under Option B to an appropriate extrapolation category (Table 8-2). For similar information for Option A and the final Option, see McGuire 2004a.

Furthermore, EPA assumed that a facility's water quality response to the regulation would be similar to other facilities in the same extrapolation category, and therefore developed case studies for key categories. Additionally, EPA assumed that each in-scope facility from the detailed survey sample represents a specific number of facilities in the total population of in-scope facilities, and that the specific number of in-scope facilities from the total population can be adequately represented by the detailed survey facility's sample weight. Table 8-2 also shows national estimates for the number of in-scope facilities for each extrapolation category. The 24 in-scope detailed questionnaire facilities that have load reductions (and for which EPA has detailed survey data) nationally represent 86 facilities with load reductions. There were 9 detailed questionnaire facilities (when multiplied by sample weights, this equals

---

[10]Median value estimate for baseline TSS concentration from median value indicated on April 2, 2004 Tetra Tech spreadsheet (Tetra Tech, 2004).

[11]Median value estimate for dilution ratio from median value indicated on April 2, 2004 Tetra Tech spreadsheet (Tetra Tech, 2004).

approximately 27 facilities in the national population of facilities affected by the final regulation) that could not be accurately categorized because of missing receiving-water flow data.

Table 8-2
Definition of Extrapolation Categories* and National Estimates for the Number
of Facilities In-scope for the Regulation—Option B Only

(A) Extrapolation Category	(B) Extrapolation Category Label	(C) Number of In-Scope Detailed Questionnaire Facilities	(D) National Number of In-scope Facilities
% TSS load reduction low = L Baseline TSS concentration low = L Dilution ratio low = L	LLL	2	7
% TSS load reduction low = L Baseline TSS concentration low = L Dilution ratio high = H	LLH	3	11
% TSS load reduction low = L Baseline TSS concentration high = H Dilution ratio low = L	LHL	1	4
% TSS load reduction low = L Baseline TSS concentration high = H Dilution ratio high = H	LHH	4	18
% TSS load reduction high = H Baseline TSS concentration low = L Dilution ratio low = L	HLL	0	0
% TSS load reduction high = H Baseline TSS concentration low = L Dilution ratio high = H	HLH	1	4
% TSS load reduction high = H Baseline TSS concentration high = H Dilution ratio low = L	HHL	2	7
% TSS load reduction high = H Baseline TSS concentration high = H Dilution ratio high = H	HHH	2	8
Missing receiving water flow data	n/a	9	27
Total	n/a	24	86

NOTE: All facilities represented in this Table are estimated to have non-zero load reductions under the Option B. For similar information for Option A and the final Option, see McGuire 2004a.
* See text for an explanation of extrapolation categories. Values in Column (D) are rounded; they are obtained by summing the sample weights associated with all facilities represented in Column (C).

8.2.4   Water Quality Modeling

8.2.4.1 Selection and Development of Case Studies

Resource and data limitations constrained the number of QUAL2E applications that could be performed. EPA developed QUAL2E models for one representative "case study" facility for the following extrapolation categories: LHL, LHH, HLH, and HHL. QUAL2E simulations were also performed for the HHH extrapolation category, using QUAL2E models already developed for the LHH category. A more detailed discussion of this process is discussed below. Case studies were not performed for the LLL, LLH, and HLL categories because (a) no facilities were in the HLL category and (b) EPA focused modeling resources on categories expected to represent a larger proportion of benefits. Water quality improvements for facilities in the LLL and LLH categories were expected to be smaller than improvements for facilities in the other categories.

Since in-scope CAAP facilities are located throughout the United States, EPA considered CAAP facilities throughout the country when choosing representative "case studies." Water quality models were developed and configured for existing, monitored facilities. Each case study model was configured with receiving stream characteristics to represent geographically similar conditions at the existing facility.

EPA's selection process of these case study sites also considered the following information:

- Availability of physical data from similar local streams for configuration of model inputs
- Availability of stream water quality data for developing upstream and downstream conditions
- Amount of data available for CAAP facility effluent flows (water quality and magnitude of flows) to accurately characterize stream inputs from the facility
- Type of CAAP facility production system and species

Availability of data for model configuration and calibration[12] was a key consideration for study site selection. Locations of water quality and streamflow monitoring stations around existing facilities were obtained from the EPA's BASINS and STORET databases and USGS. In addition, BASINS datasets were utilized for identification of environmental and spatial features of the receiving stream. The final selection of sites for developing water quality models represented a balance between available resources, the accessibility of the suitable data, and the number of facilities that could be represented with a specific site.

For each selected study site, background information was collected regarding characteristics of the watershed, stream, and CAAP facility. This information included analyses of the physical extent of the watershed to the point of the CAAP facility's discharge, land use within the watershed that are potential nonpoint sources of pollution to the stream, proximity of other dischargers that could potentially influence analysis of the isolated impact of the CAAP facility, and other environmental or meteorological attributes of the region that distinguish the study site. Additional information about how the specific sites were selected is available (Hochheimer, 2004).

---

[12] EPA performed calibration adjustments to many of the QUAL2E model input variables during the modeling process. See the model reports (Hochheimer et al., 2004 a-d) for more information on model parameterization.

Special attention was placed on selection of appropriate study sites to ensure that hydraulic/hydrologic and loading processes were not present that would affect model configuration and calibration, and analysis of model results. An ideal study site will attempt to isolate the impacts of a CAAP facility so that modeling analysis will not be impaired by unforeseen influences not simulated by the model.

8.2.4.2 Model Configuration

The selected model for analysis of CAAP facility impacts during proposal was a steady state application of QUAL2E (Brown and Barnwell, 1987). EPA extended this application to model CAAP facilities using QUAL2E during critical design conditions (e.g., low flow, high temperatures) at specific facilities that provided information in the detailed survey. QUAL2E is capable of simulating up to 15 water quality constituents, including:

- Dissolved oxygen
- Biochemical oxygen demand
- Temperature
- Chlorophyll a
- Organic nitrogen
- Ammonia
- Nitrite
- Nitrate
- Organic phosphorus
- Dissolved phosphorus
- Coliforms
- An arbitrary nonconservative constituent
- 3 conservative constituents

Relative to other models currently available, QUAL2E was selected as the ideal tool for impact analysis due to input data requirements and parameters modeled. QUAL2E provided EPA with the ability to simulate several constituents using model processes that can be logically parameterized and justified using assumptions based on either collected data or literature. Moreover, the detail of the processes modeled by QUAL2E provided EPA with a good balance between available data, assumptions required, and ability to validate to observed data so that model configuration can be refined. For example, little data is generally available to describe sediment oxygen demand (SOD) in most streams, although this process is a key component in prediction of in-stream water quality. QUAL2E allows designation of a zero-order SOD term that can be refined through the validation process.

To configure the hydraulic characteristics of the streams for the model, EPA reviewed available physical data and literature values for parameterization of hydraulic equations utilized by QUAL2E. To configure the physical attributes of the streams, EPA estimated stream cross-sections by using one of these methods: 1) observed data (e.g., USGS stream gages) for the study site, 2) data from a neighboring stream with similar hydraulic characteristics, or 3) empirical methods. EPA estimated longitudinal profiles from digital elevation model (DEM) data. EPA refined the hydraulic model configuration by comparing model-predicted flows to observed data.

For configuration of the steady-state model of each study site, QUAL2E requires the assumption of constant flows and water quality for each model input including all point sources (CAAP facility), upstream flow, and inflow from tributaries or groundwater. For CAAP facility inflow, EPA determined flow magnitude and water quality from average observed conditions reported in the detailed survey. EPA assessed critical conditions in the stream by adjusting the model conditions to particular seasons or periods when stream impacts are of maximum concern (e.g., summer low flow period or period of maximum feeding). Specifically, EPA derived critical low (7Q10) estimates for months of high facility production levels and used these flows and production levels to drive the QUAL2E simulations. For background flows and water quality (upstream, tributary, and groundwater), EPA estimated values from observed data. When data was limited to describe background conditions, EPA collected data from similar neighboring streams. EPA carefully selected study sites with plentiful stream data to reduce the assumptions required to address such data gaps.

EPA configured water quality processes utilized by QUAL2E by using literature values. Such processes included mass transport (including first-order decay and settling), sediment oxygen demand (user-specified rate), algal growth as a function of temperature (via solar radiation), and algal, nitrogen, phosphorus, and dissolved oxygen interactions.

It is important to stress that the inputs for the case studies were synthesized from several data sources and the case studies themselves should not be considered realistic representations of specific regulated facilities. For example, EPA used water quality data from not only a single local sampling station or stream, but also considered data from similar streams in the watershed to develop more robust estimates of background conditions of the receiving stream at the point of CAAP discharge. EPA also in some cases used flow data from nearby watersheds and used watershed size to extrapolate flow data on the subject stream when monitoring data was not available (e.g., the data was not recent, flow data was not recorded at the gage). Rather, the water quality modeling case studies were used to develop a relationship between the key factors driving water quality response (percent TSS load reduction, baseline TSS effluent concentration, and dilution ratio) and simulated water quality response. The simulated water quality response for any given case study was then assumed to be valid for all facilities in the scope of EPA's final regulation with similar values for percent TSS load reduction, baseline TSS concentration, and dilution ratio (i.e., in the same "extrapolation category").

The following briefly summarizes basic facility and geographic information about the case studies EPA evaluated.

Case Study 1 ("LHL")

A case study to represent the "LHL" extrapolation category was developed using a facility located in the Blue Ridge Ecoregion (Central and Eastern Forested Uplands Nutrient Ecoregion) in the southeastern United States. Consistent with the definition of this extrapolation category (see earlier discussion), this facility has a "Low" regulatory percent TSS load reduction, a "High" baseline TSS concentration, and a "Low" dilution ratio. See Table 8-3 for specific values for this case study.

This government-owned facility uses a flow-through system to produce over 100,000 pounds of trout each year. The watershed encompasses over 2,000 square miles[13]. Primary land uses for this watershed include forestry and grazing[14]. Average temperature for this region is approximately 60 °F and average annual precipitation is approximately 50 inches[15]. A number of other detailed questionnaire facilities are located in the same general area.

Case Study #2 ("HLH")

A case study to represent the "HLH" extrapolation category was developed using a facility located in the Pacific Northwest. Consistent with the definition of the HLH extrapolation category, this facility has a "High" regulatory percent TSS load reduction, a "Low" baseline TSS concentration, and a "High" dilution ratio. See Table 8-3 for specific values. The selected facility is a government-owned salmon facility that is located in Coast Range Ecoregion (Western Forested Mountains Nutrient Ecoregion) of the United States. This facility produces just under 100,000 pounds of salmon annually. The watershed encompasses just over 670 square miles[16]. Average annual temperature is 51° F, and the mean annual precipitation is approximately 67 inches[17]. A number of other detailed questionnaire facilities are located in the same general area.

Case Study #3 ("LHH")

A case study to represent the "LHH" extrapolation category was developed using a facility located in the upper Midwest. Consistent with the definition of the LHH extrapolation category, this facility has a "Low" regulatory percent TSS load reduction, a "High" baseline TSS concentration, and a "High" dilution ratio. See Table 8-3 for specific values.

The selected facility is a government-owned trout facility that is located in the Northern Lakes and Forests Ecoregion (Nutrient Poor Largely Glaciated Upper Midwest and Northeast Nutrient Ecoregion) of the United States. The facility reports an annual production of over 200,000 pounds of trout. The watershed, which is approximately 1,600 square miles[18], has forestry as its primary land use,

---

[13] Environmental Statistics Group (ESG) provides several sources of watershed size. Available online at *http://www.esg.montana.edu*.

[14] Conservative Technology Information Center, Purdue University.

[15] Climatic data was obtained from NOAA. Since it was not available for the exact facility location, data from nearby were used to approximate conditions at the facility.

[16] Environmental Statistics Group (ESG) provides several sources of watershed size. Available online at *http://www.esg.montana.edu*.

[17] Climatic data was obtained from NOAA.

[18] Environmental Statistics Group (ESG) provides several sources of watershed size. Available online at *http://www.esg.montana.edu*.

with cropland and grazing as secondary uses[19]. The annual high temperature in this area is 80° F, with an average low temperature of 10° F, and average annual precipitation of 34 inches[20]. A number of other detailed questionnaire facilities are located in the same general area.

Case Study #4 ("HHL")

A case study to represent the "HHL" extrapolation category was developed using a facility in California. Consistent with the definition of the HLL extrapolation category, this facility has a "High" regulatory percent TSS load reduction, a "High" baseline TSS concentration, and a "Low" dilution ratio. See Table 8-3 for specific values.

The selected facility is a government-owned trout facility that is located in the Southern and Central California Chaparral and Oak Woodlands Ecoregion (Xeric West Nutrient Ecoregion) in the United States. The facility produces over 400,000 pounds of trout annually. The watershed is over 800 square miles[21]. The mean annual temperature is approximately 61° F, with mean annual rainfall of approximately 33 inches[22]. A number of other detailed questionnaire facilities are located in the same general area.

Estimating Benefits from Extrapolation Categories Not Modeled as Case Studies

EPA explored estimating benefits from the remaining extrapolation categories not already modeled with QUAL2E. As stated before, HLL was not considered because there are no facilities in this extrapolation category. Of the remaining categories (HHH, LLL, and LLH), EPA chose to estimate benefits from the HHH category because it had the highest percent TSS load reduction of the three categories and water quality improvements for facilities in the LLL and LLH categories were expected to be smaller than improvements from facilities in the HHH category. EPA first estimated benefits of the HHH category by using the QUAL2E model already developed for Case Study #2 (HLH). To accomplish this, EPA chose one representative facility from the HHH category (the facility from the nine HHH facilities whose dilution ratio is closest to the dilution ratio for Case Study #2) and adjusted the facility flow to match the flow of the receiving water in the Pacific Northwest. The effluent concentrations were adjusted accordingly). EPA then applied this facility effluent data to the model for Case Study #2. To test the sensitivity of using a different Case Study model to simulate water quality improvements from the HHH extrapolation category, EPA also used the model for Case Study #3. Larger water quality improvements were observed for the adjusted Case Study #2, in comparison to the simulated water

---

[19] Conservative Technology Information Center, Purdue University.

[20] Climatic data from the county level was used since data were not available for the exact location of the facility. Data obtained from NOAA was reported incorrectly, so EPA obtained corrected data from the county.

[21] Environmental Statistics Group (ESG) provides several sources of watershed size. Available online at *http://www.esg.montana.edu.*

[22] Since climate data were not available at the exact location of the facility, data from the nearest NOAA monitoring station were used.

quality improvements for adjusted Case Study #3. Therefore, EPA carried out "HHH" benefits monetization on the results for adjusted Case Study #3 to avoid overestimating benefits of the CAAP rule.

8.2.4.3 Model Results

For each of the case studies described above, including the HHH simulation, QUAL2E was used to generate simulated concentrations for selected water quality parameters over a 30 km distance downstream of each facility under both baseline and post-regulatory loading scenarios. QUAL2E output for DO, BOD, TSS, NO_3, and PO_4 are used to estimate a "water quality index" (WQI6) value, which in turn is used to estimate monetized benefits of improvements to water quality. The specific form of the function relating these water quality parameters to WQI6 is described in Section 8.2.4 of this Chapter.

Figure 8-1 displays example QUAL2E output for the BASELINE scenario at the QUAL2E simulation for Case Study #3, as adjusted to represent the "HHH" extrapolation category (documentation of all QUAL2E runs can be found in the Record for this rulemaking (Hochheimer et al., 2004 a-d)). Increases in pollutant concentrations can be seen a short distance downstream of the facility, located at river kilometer (RK) 0.5. Simulated improvements in water quality following regulatory loading reductions can be seen in Figure 8-2. Peak pollutant concentrations are lower following the regulation, resulting in a small increase in the value of WQI6 (Figure 8-3). The monetized benefit of the upward shift in the value of WQI6, best seen on Figure 8-3, is calculated as described in the following section.

EPA performed limited calibration on the case study models in the form of adjustments to input parameters, which were necessary to achieve reasonable values for the results. Because EPA was primarily interested in monetizing the benefits associated with regulatory changes, analysis of the relative differences in stream water quality based on changes to facility loads before and after regulation was most important, and EPA sought to calibrate the model so it could generate reasonable changes in water quality (rather than to calibrate the model to achieve accurate, absolute values for the water quality parameters). Therefore, EPA focused its calibration to ensure that the model output values were within normally expected ranges of values for the water quality parameters of interest. In the calibration, EPA adjusted model inputs that affect processes in the stream or contributing watersheds including the BOD_5 coefficient and coefficients for nitrogen, phosphorus and algae, such as oxygen-nitrogen hydrolysis and ammonia oxidation, which were important to ensure that the model represents streams similar to those located adjacent to the case study facilities.

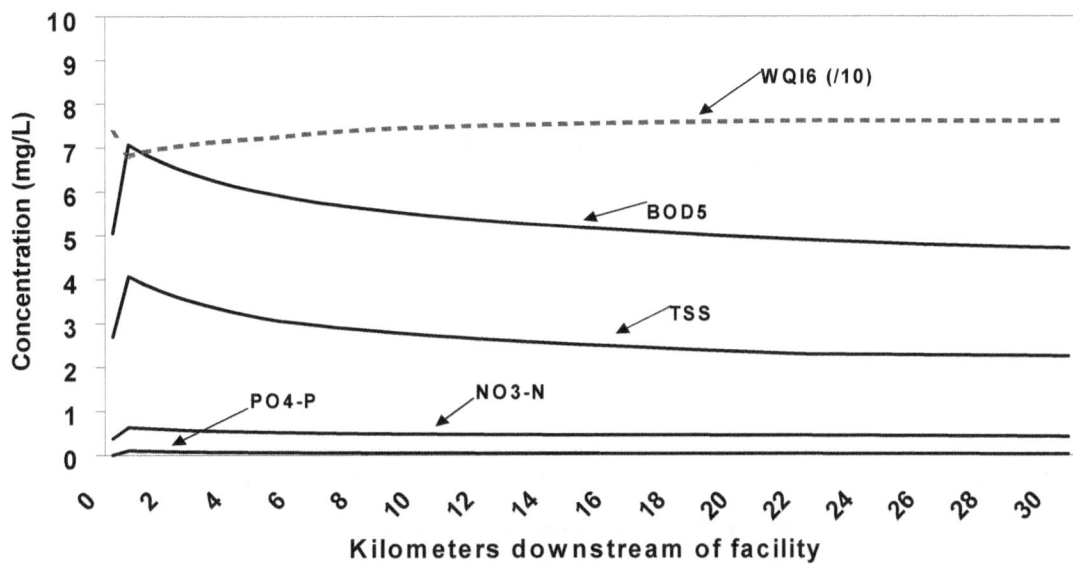

Figure 8-1. Sample QUAL2E output for ***baseline*** discharges in a 30 km reach downstream of a case study facility. The facility is located at river kilometer 0.5. Simulated BOD5, TSS, NO3-N, and PO4-P are shown by solid lines; aggregate water quality index (WQI6) - a function of the simulated parameters - is shown by the broken line. WQI6 divided by 10 to enable display on the same graph.

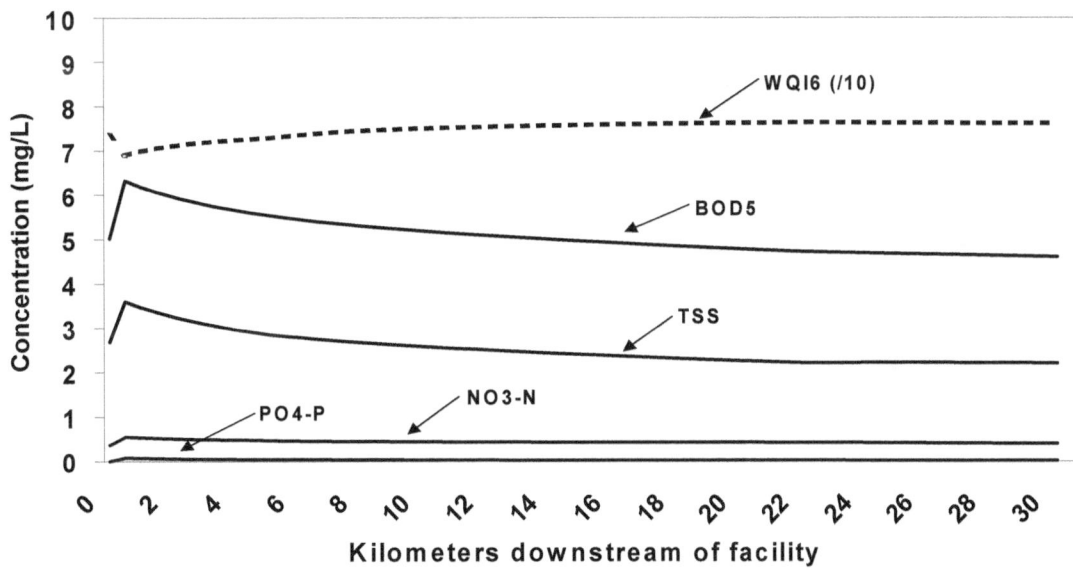

Figure 8-2. Sample QUAL2E output for ***post-regulatory*** discharges in the same reach and facility as shown in Figure 1. As in Figure 1, facility is at RK 0.5. See caption for Figure 1 and text for further discussion.

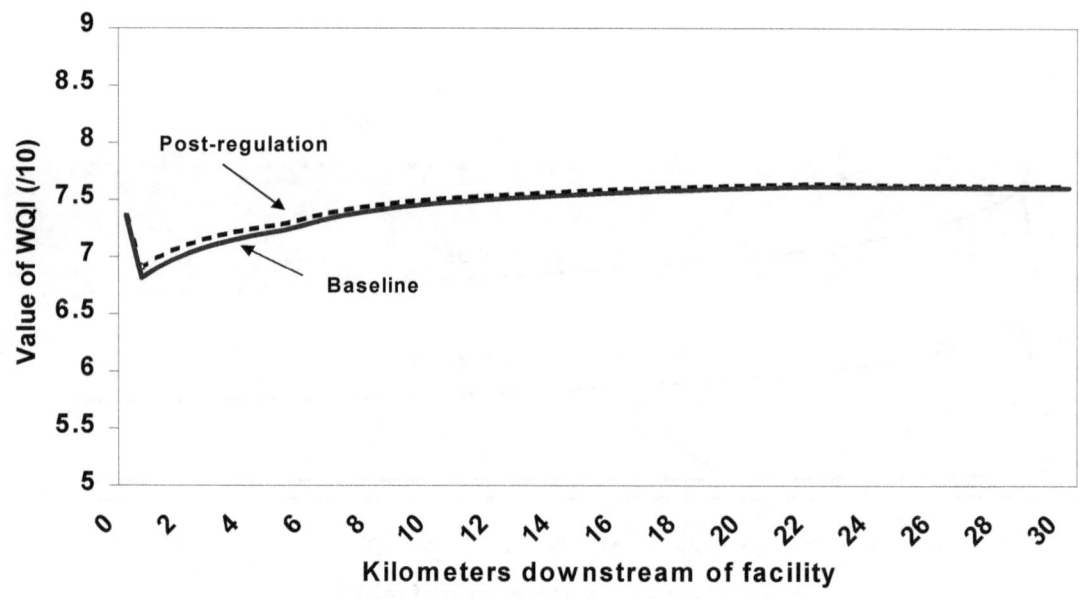

Figure 8-3. Comparison of baseline and post-regulatory WQI6 values from sample QUAL2E output presented in Figures 1 and 2. As in Figures 1 and 2, facility is located at RK 0.5. Upward shift in WQI6 indicates improved water quality. Monetized benefits at each facility are based on the cumulative improvement in water quality along the length of the 30 km simulated reach. See text for further explanation.

The following table summarizes QUAL2E model run characteristics and results:

Table 8-3
Summary of QUAL2E run results.

(A) Name/description	(B) % TSS load reduction	(C) Baseline TSS effluent concentration (mg/L)	(D) Dilution ratio	(E) Simulated change in water quality* (reach-average ΔWQI6)
Case study 1 ("LHL") - facility in Blue Ridge ecoregion	1.2	10.01	0.17	0.0049
Case study 2 ("HLH") - facility in Coast Range ecoregion	21.3	0.63	0.52	0.0385
Case study 3 ("LHH") - facility in Northern Lakes and Forest ecoregion	3	3.63	0.28	0.0109
Case study 4 ("HHL") - facility in Southern and Central CA Chaparral and Oak Woodlands ecoregion	51.6	1.5	0.18	0.2642
Case study 3, modified to represent "HHH" extrapolation category	21.9	13.1	0.37	0.3325

*Values in Column (E) from results reported in Miller, 2004.

8.2.5   Economic Valuation

8.2.5.1 Economic Valuation Approach

The process for assigning a dollar value to changes in water quality for each sample case study affected by the CAAP rule involves the following steps: (1) calculate changes in aggregate water quality index (WQI) values, based on predicted changes in water quality parameter concentrations, and (2) estimation of household willingness to pay (WTP) for the change in WQI, and (3) summation of benefits based on in-State and out-of-State populations of households.

In the first step, simulated water quality parameter changes for each case study and the HHH simulation are translated into a composite water quality index (WQI) value. The original WQI, from which the WQI used for CAAP rule was derived, included nine water quality parameters: five-day biochemical oxygen demand (BOD_5), percent dissolved oxygen saturation (% DOsat), fecal coliform bacteria (FEC), total solids (TS), nitrate (NO_3), phosphate (PO_4), temperature, turbidity, and pH. The concentrations of each water quality parameter are mapped onto a corresponding index number between 0 and 100 (zero equating with poor water quality) using functional relationship curves (McClelland, 1974). McClelland derived the functional relationships by averaging the judgments from 142 water quality experts. A composite WQI is estimated using the parameter specific weights and the function below; weights are again, based on the summary judgments of the expert panel.

$$\text{Composite WQI} = \prod_{i}^{I} X_i^{\alpha_i}, \quad \sum_{i} \alpha_i = 1$$

where,

I = number of water quality parameters in the composite index,
X = index value for individual water quality parameter (0 to 100), and
α = parameter-specific weight.

For previous rulemakings, load reduction data, water quality data, and/or modeling capability did not extend to all nine parameters, so a modified WQI formulation had been developed for four of the parameters (WQI-4). The parameters were dissolved oxygen (DO), biochemical oxygen demand (BOD), total suspended solids (TSS), and fecal coliform (FC). EPA applied this version of the WQI for the proposed rule. Because we do not expect loadings for FC to be discharged from CAAP facilities, we assumed that background levels of this parameter remain unchanged. EPA adopted a six-parameter WQI (WQI-6) for the final CAAP rule based on TSS, BOD, DO, FC, plus nitrate (NO_3) and phosphate (PO_4). The new index more completely reflects the type of water quality changes that will result from loading reductions for TSS, total nitrogen (TN), total phosphorus (TP), and BOD. Final rule benefits presented here were calculated on the basis of WQI-6. In the original index, McClelland (1974) used turbidity in her assessment rather than TSS. To incorporate TSS in the analysis for the final CAAP rule, a regression equation is therefore used to convert the original functional relationship curve of water quality against turbidity into a curve of water quality against TSS. The weight on each parameter was also recalculated so that the sum of weights equals one, thereby insuring that the composite index continued on a 0-100 basis even though it had fewer components. For the benefits analysis for the final rule, WQI-6 values are estimated before (i.e., baseline) and after implementation of final CAAP rule requirements for each half kilometer increment of the total 30 kilometer stream reach distance modeled for each case study and the HHH simulation.

In the second step, household willingness to pay (WTP) values are estimated for changes in WQI-6. Economic research indicates that the public is willing to pay for improvements in water quality and several methods such as stated preference surveys have been developed to translate changes in water quality to monetized values. At proposal, EPA based the water quality benefits monetization on household WTP values for discrete changes in recreational use classifications (e.g., boatable to fishable, fishable to swimmable water quality) as derived from a stated-preference survey conducted by Carson and Mitchell (1993). EPA divided the willingness-to-pay (WTP) values for changes in recreational water "use classes" by the number of WQI units associated with each use class. For example, Carson and Mitchell's survey informed the respondent that boatable, fishable, and swimmable waters are mapped onto respective ranges of WQI values of 25 to 50, 50 to 70, and greater than 70. EPA was therefore able to assign incremental WTP values for each unit change in the aggregate WQI.

For the final CAAP rule, EPA adopts an alternative approach, also based on Carson and Mitchell's work. In addition to describing their results in the form of WTP for discrete changes in recreational use classifications, the authors also estimated household WTP as a function of the WQI representing all of the nations waters and household income.

Carson and Mitchell (1993) derived an equation to assess the value of water quality along the continuous WQI scale using the responses to their national survey. Assuming that the proportion of families engaging in water-based recreation and the proportion of respondents who feel a national goal of protecting nature and controlling pollution is very important are the same as when the Carson and Mitchell survey was completed, the incremental value associated with increasing WQI from WQI_0 to WQI_1 can be calculated as:

$$WTP_{TOT} = \exp[0.8341 + 0.819\log(\tfrac{WQI_1}{10}) + 0.959\log(\tfrac{Y}{1000}) -$$
$$\exp[0.8341 + 0.819\log(\tfrac{WQI_0}{10}) + 0.959\log(\tfrac{Y}{1000})]$$

where

WTP_{TOT} = a household's willingness-to-pay for increasing water quality (1983 dollars)
Y = household income (sample average = \$35,366 in 1983 dollars)
WQI_1 = Composite water quality index under regulatory scenario
WQI_0 = Composite water quality index under baseline

In this case, Y was selected to correspond to an estimated mean household income of \$35,366 in 2003 expressed as 1983 dollars (note: 2003 mean household income projected using US Census 2001 mean household income and percent increase in Bureau of Economic Analysis real per capita disposable income from 2001 to 2003; 2003 income adjusted to 1983 dollars using CPI-U-RS). The resulting value estimates were inflated to 2003 dollars using the growth rate in the consumer price index (CPI) of 1.8574 since 1983 (U.S. Department of Labor, Bureau of Labor Statistics, www.bls.gov/cpi). WTP_{TOT} values are estimated for each change in WQI for each half kilometer increment of each 30 kilometer in the models. The sum of values for the modeled reach is equal to the monetized value for a single household.

In the third step, EPA estimates benefits for the total population of households. Benefits are calculated state-by-state and are broken down into local and non-local benefits. Carson and Mitchell (1993) found that respondents were willing to pay more for water quality improvements within their own state, and estimated that 2/3 of the total willingness-to-pay applied to in-State water quality changes. Non-local benefits correspond to the amount a population is willing to pay for water quality improvements outside of their own state, and were estimated as 1/3 of the total willingness-to-pay (i.e., it assumes households will allocate two-thirds of their willingness to pay to improvements in-State waters). For details about final benefit calculations, see Miller (2004).

8.2.5.2 Uncertainties and Other Considerations Regarding Benefits Valuation

As noted above, EPA relies on a willingness to pay function derived by Carson and Mitchell to value changes in the water quality index for reaches affected by this rule. This function has the ability to capture benefits of marginal changes in water quality. Based on this approach, EPA is able to assess the value of improvements in water quality along the continuous 0 to 100 point scale, and values are less sensitive to the baseline use of the water body (relative to methods used for the proposed rule). The calculation of benefits is completed separately for each State and takes into account differences in willingness to pay for local and non-local water quality improvements. Note that the WTP function assumes decreasing marginal benefits with respect to water quality index values; this is consistent with consumer demand theory and implies that willingness to pay for incremental changes in water quality decreases as index values increase. There are a number of other issues associated with the transfer of values from the Carson and Mitchell survey results that affect benefit estimates for this final rule, and these issues are discussed below.

Economic benefits of the this rule can be broadly defined according to categories of goods and services provided by improved water quality: use and nonuse benefits. The first category includes benefits that pertain to the use (direct or indirect) of the affected resources (recreational fishing). The direct use benefits can be further categorized according to whether or not affected goods and services are traded in the market (commercial fish harvests). For this rule, EPA has not identified any goods that are traded. The non-traded or non-market "use" benefits implicitly assessed in this final rule include recreational activities. Nonuse benefits occur when environmental improvements affect a person's value for a natural resource that is independent of that person's present use of the resource. Nonuse values derive from people's desire to bequeath resources to future generations, vicarious consumption through others, a sense of stewardship or responsibility for preserving ecological resources, and the simple knowledge that a resource exists in an improved state.

When estimating nonuse benefits, it is not possible to directly observe people using the good or resource, therefore, more traditional revealed preferences economic methods such as travel costs are not applicable to the derivation of nonuse values. Instead, analysts survey people and directly ask them to state their preferences or willingness to pay for an environmental improvement (e.g., what are you willing to pay to improve water quality from boatable to swimmable). Statistical models are used to compile these survey responses and derive nonuse values for the resource improvements specified in the survey questions[23]. The values estimated from stated preference surveys may capture both use and nonuse values depending on how the survey is implemented.

The Carson and Mitchell stated preference study is a case were both use and nonuse benefits were estimated (i.e., Total willingness to pay). The willingness to pay values developed in their national survey are the basis for the benefits transfer, which produced the total benefit values sited in this report. Carson and Mitchell asked respondents how they would divide their total willingness to pay values for improved water quality between their home state and the rest of the nation. The fact that Carson and Mitchell were asking people to value significant changes in water quality across the nation can present a source of error in the estimation of the benefits for today's rule. This is due to the imprecise fit between the scenario presented in their survey questions and the more narrow scope, both in terms of the number of water bodies and the size of the water quality change, of the CAAP rule. The direction of the impact produced by this difference between the survey and policy scenarios on our estimated use and nonuse benefits, for today's rule, is unclear.

EPA notes that an additional source of indeterminate error is imposed by the benefits transfer framework stemming from the assumption that willingness to pay for the same level of water quality

---

[23]In 1993, the National Oceanic and Atmospheric Administration (NOAA) convened a panel of economists to evaluate a form of stated preference methods (*contingent valuation* (CV)) and to devise a set of "best practices" for designing and implementing CV surveys. The NOAA recommendations are in the Federal Register (1994). EPA has subsequently published "considerations in evaluating CV studies" and discusses other stated preference methods in the agency's *Guidelines for Preparing Economic Analyses* (2000). OMB's most recent draft of "best practices" for conducting regulatory analysis, recognizes nonuse values and provides guarded acceptance of stated preference methods by listing "principles that should be considered" when evaluating the quality of such a study (Draft OMB Circular A-4, 9/17/03).

improvements, from the same baseline level of quality, are constant across all water-bodies. This restriction implies that people have the same value for a similar improvement in water quality in water bodies that may differ in terms of geographic location, surrounding land use, and recreational use pressure.

Two additional sources of error can be identified that would tend to produce an underestimate of use and nonuse benefits for the rule. Values returned by stated preference studies are sensitive to the language used to inform respondents about the baseline conditions and the changes in resource produced by the policy being evaluated. The nonuse component of Carson and Mitchell's reported total willingness to pay may be under estimated because of the use of recreational activity based titles for differing water quality categories i.e. boatable, fishable, swimmable. These designations are likely to produce cognitive links in respondent's minds to benefits associated with recreational uses, and down play the role of nonuse benefits. Recreational "tags" may have lead to an incomplete recognition of nonuse benefits in Carson and Mitchell's total willingness to pay valuation and therefore under-estimation of benefits for the rule.

An issue in applying the results of the Carson and Mitchell survey in the context of the water quality index is the treatment of water quality changes occurring below the boatable range and above the swimmable range. There are concerns that the survey's description of non-boatable conditions (i.e., index values less than 25) was exaggerated (i.e., unsafe for boating and swimming and unfishable), which implies that willingness-to-pay estimates for improving water to boatable conditions (i.e., index increases above 25) may be biased upwards. The survey did not ask respondents how much they would be willing to pay for improved water quality above the swimmable level.[24] These issues increase the uncertainty associated with valuing water quality changes outside the boatable to swimmable range (i.e., for water quality index values below 26 or above 70). In recognition of this uncertainty, EPA determined that some percentage of the benefits are derived from changes in water quality outside the boatable to swimmable range (i.e., less than 25 or greater than 70).

In addition to the valuation function, there is also uncertainty associated with the water quality index. The water quality index used in monetization for the final rule relies on judgements of water quality experts from the 1970s when they were asked to assign index values to different levels of individual pollutant parameters. There is some evidence suggesting that updating index values may be appropriate. This can be illustrated through a discussion of the nutrient values in the index in comparison to recent work on nutrient criteria development. EPA's recently recommended section 304(a) ecoregional water quality criteria for nutrients to define reference conditions for reducing and preventing cultural eutrophication. Index values for nitrate nitrogen and phosphate phosphorus nutrient criteria representing 304(a) 50th percentile (i.e., median) reference conditions of 'least impacted' streams are relatively high as indicated in Table 8-4. Given that fishable water quality is designated as starting at an index value of 50, swimmable at 70, and water quality suitable for drinking without treatment at 95, these results suggest that the index is overestimating baseline water quality index values associated with nutrients (e.g., 50% reference conditions for healthy aquatic life are an average of 92 and 93 index units for PO_4 and NO_3 respectively, well above an assumed index value of 50 for fishable water). Overestimation of baseline index values potentially translates into underestimation of benefits given that marginal willingness to pay for incremental changes in water quality decreases as baseline water quality increases (i.e., demand decreases with quantity). This result may be offset to some extent by the possibility that modeled

---

[24] However, respondents were made aware of the potential for water quality to improve beyond swimmable in the ladder (e.g., drinkable).

changes in nutrient concentrations will be translated into small changes in index value as the nonlinear index curve becomes more convex. In general, these results suggest that the water quality index may not reflect current evidence about the contribution of nutrients to water quality, as represented by recent 304(a) recommended ecoregional water quality criteria for nutrients. While this discussion has focused on nutrients, similar issues may be applicable to index values for TSS and BOD.

Table 8-4
Index Values for Nutrient Criteria

50% Reference Conditions[1]		Estimated 50% Criteria[2]		Parameter Index Values[3]	
Total P	Total N	PO4-P	NO3-N	PO4-P	NO3-N
0.07 mg/l	1.1 mg/l	0.053 mg/l	0.97 mg/l	92	93

1. Average of section 304(a) ecoregion water quality criteria representing 50th percentile reference conditions of 'least impacted' streams across 14 ecoregions.
2. Estimated criteria derived from 50% reference conditions and the following relationships [PO4-P] = 0.75*[TP], [NO3-N] = 0.9*[TN]
3. Index values derived by inserting 'Estimated 50% Criteria into regression functions fitted to index curves for PO4-P and NO3-N from McClelland (1974) (i.e., index curves 'map' concentrations into index values).

8.2.6a   Estimated National Water Quality Benefits—Options A and B

EPA estimates the national water quality benefits of Option A to be $84,000 or, for Option B, $94,000 to $118,000. Table 8-5a summarizes data used to develop the national benefit estimate for Option B. As described in Section 8.2.2, each of the 24 in-scope, detailed questionnaire facilities with non-zero load reductions were assigned to an appropriate extrapolation category where data allowed (Column (C) of Table 8-5a). Each facility was further assigned the value of change in WQI6 (d(WQI6)) corresponding to the appropriate extrapolation category (Column (F) in Table 8-5a; see also Table 8-3). As noted earlier in this Chapter, receiving water flow data, and thus a complete extrapolation categorization, could not be done for 9 of the 24 facilities in Table 8-5a.

As described in Section 8.2.4.1, monetized benefits for d(WQI6) are calculated on a state-by-state basis. Column (B) of Table 8-5a indicates the EPA region in which each facility is located (the State in which the facility is located was used in monetizing benefits, but EPA region, rather than State, is provided in Column (B) as a means of aggregation to protect potential confidential business information (CBI)). Thus, the value in Column (G) indicate the monetized benefit calculated for the appropriate d(WQI6), taking into consideration the State in which the facility is located. Column (H) represents the benefit for all facilities in the national, regulated population, represented by the detailed questionnaire facility (i.e., the sample weight for the detailed questionnaire facility, Column (D), multiplied by the benefit value for the detailed questionnaire facility in Column (G)).

Two additional steps were taken to estimate the water quality benefits. First, EPA assumes that it is more appropriate to apply the Carson-Mitchell valuation method to larger-sized streams, where recreation is more probable, rather than smaller-sized streams where recreation is less probable. Accordingly, EPA took the step of omitting from the monetized benefit analysis certain facilities located on smaller-sized streams. To do this, for all facilities for which receiving water flow data could be found, EPA determined whether the stream was part of a national subset of larger streams (referred to as

Table 8-5a
National Water Quality Benefit Estimate for Option B

(A) No.	(B) EPA Region	(C) Extrap. Category	(D) Sample Weight	(E) RF3Lite Flag	(F) d(WQI6)	(G) Estimated Benefit	(H) Extrap. Benefit (L)	(I) Extrap. Benefit (H)
1		HHH	5.35	0	0.3325			
2		HHH	3.73	1	0.3325	$ 16,741	$ 62,505	$ 62,505
3		HH_	1.31	n.d.	n.d.			$ 16,423
4		HHL	3.87	0	0.2642			
5		HHL	3.92	1	0.2642	$ 5,537	$ 21,683	$ 21,683
6		HLH	3.73	1	0.0385	$ 1,919	$ 7,165	$ 7,165
7		HL_	1.53	n.d.	n.d.			$ 4,477
8		LHH	3.67	0	0.0109			
9		LHH	3.89	0	0.0109			
10		LHH	5.22	1	0.0109	$ 237	$ 1,238	$ 1,238
11		LH_	4.66	n.d.	n.d.			$ 1,440
12		LH_	4.10	n.d.	n.d.			$ 427
13		LHH	3.70	1	0.0109	$ 327	$ 1,211	$ 1,211
14		LH_	3.49	n.d.	n.d.			$ 1,266
15		LHL	1.78	0	0.0049			
16		LH_	3.68	n.d.	n.d.			$ 298
17		LH_	1.39	n.d.	n.d.			$ 140
18		LLH	3.73	0	n.d.			
19		LLH	3.73	1	n.d.			
20		LL_	1.37	n.d.	n.d.			
21		LL_	3.46	n.d.	n.d.			
22		LLH	3.68	1	n.d.			
23		LLL	3.68	1	n.d.			
24		LLL	3.68	1	n.d.			
						Option B *TOTAL*	$ 93,803	$ 118,274

the RF3 Lite network of streams[25]. A value of 1 in Column (E) indicates that a facility is part of the RF3 Lite network. EPA did not estimate any benefits for facilities determined not to be part of the RF3 Lite network (i.e., with a value of 0 in Column (E).) Thus, if the value in Column (E) is 0, then there are no benefit values in Columns (G), (H), or (I).

Second, for facilities where receiving water flow data (and thus dilution ratio) could not be determined (i.e., facilities with "n.d." in Column (E)), EPA pursued two alternative assumptions. For a lower bound estimate of benefits, EPA assigned a benefit value of 0 to all facilities for which receiving water flow data could not be determined. The sum of the values in Column (H) represents this lower-bound estimate ($94K). For an upper-bound estimate of benefits, EPA essentially assumed an "average" dilution ratio value for facilities with no receiving water data and developed a benefit estimate based on this assumption. Column (I) contains the estimated benefits for these facilities, and the sum of the values in Column (I) represents this upper-bound estimate ($118K). Again, the approach for estimating the national benefit for Option A was done in the same manner as that for Option B discussed above. For more detailed descriptions of the method described in this section, see McGuire, 2004a.

8.2.6b   Estimated National Water Quality Benefits—Final Option

EPA estimates the national water quality benefits of the final Option to range from $66,214 - $98,616 (Table 8-5b). Table 8-5b indicates benefits, by extrapolation category. In general, the benefits estimate was developed using the same steps described in section 8.2.6a although the Table 8-5b is less detailed than Table 8-5a and presents only a summary of the results of these steps. The benefit estimate for the final Option also reflects minor updates to the final list of in-scope facilities and sample weights that were not reflected in the estimate for Options A and B. Please see McGuire (2004) for more detail on the calculations for the final Option.

Table 8-5b
National Water Quality Benefit Estimate for the Final Option

Extrapolation category	Total national benefit for category ($2003)
LLL-LLH	not estimated
LHL-LHH	$2,126 - $5,330
HLL-HLH	$6,591 - $12,031
HHL-HHH	$57,497 - $81,255
TOTAL BENEFITS, FINAL OPTION	$66,214 - $98-616

---

[25]The EPA Reach Files (RFs) are a series of hydrologic databases that contain information on the U.S. surface waters. The RF3 Lite subset of surface waters contains streams that are greater than 10 miles in length (and also small streams needed to connect streams greater than 10 miles in length into a complete network). There are approximately three times as many miles of streams in the overall network as compared with the RF3 Lite subset (USEPA, 2003).

8.2.7   Sources of Uncertainty

In addition to the sources of uncertainty associated with the monetization method, described in Section 8.2.4.2, there are a number of other sources that contribute uncertainty in the above estimate of water quality benefits of the regulation. Several important sources include the following:

- Uncertainty in choice of factors that drive water quality. EPA has assumed that facilities with similar regulatory percent TSS load reductions, baseline TSS concentrations, and dilution ratios (see Section 8.2.2) will experience similar changes in water quality. On the basis of this assumption, EPA assumes that all detailed questionnaire facilities in the same "extrapolation categories" may be grouped together and a single QUAL2E case study model for a facility within the category will be representative, in terms of change in WQI6, of all facilities in the extrapolation category. Errors in this assumption—WQI6 response is unrelated to these three factors—could lead to incorrectly attributing large or small water quality responses to facilities. Errors in this assumption could lead to underestimates or overestimates of benefits. As noted earlier, EPA informally evaluated the relationship between percent TSS load reduction, baseline TSS concentration, dilution ratio, and simulated water quality response. EPA performed multiple regression analyses between the three explanatory factors and change in WQI6 using four different model specifications (linear without constant, linear with constant, semilog, and log-log). The three explanatory factors explained from 73% to 99% of the variation in d(WQI6), depending on model specification. See McGuire (2004c). The regression results support the assumption that these three factors are important determinants of water quality response.

- Coarseness of extrapolation categorization. The coarseness of categorization for each factor (only two possible values, "Low" and "High," for each of the three factors) may introduce uncertainty in the benefits estimate. A larger number of extrapolation categories for each factor, or alternatively a different approach (e.g., developing a relationship between QUAL2E-simulated d(WQI) and key explanatory factors such as percent TSS load reduction, baseline TSS effluent concentration, and dilution ratio), could potentially reduce, and better enable a characterization of, this source of uncertainty.

- Uncertainty in case study specification. EPA configured each case study using EPA estimates of baseline pollutant loads, regulatory load reductions, receiving water flow and quality, and facility effluent flow. Each of these estimates is subject to some uncertainty. For example, the accuracy of reported feed use, facility flow rates, annual production, and estimates of feed conversion ratios could lead to underestimates or overestimates of both baseline and regulated load estimates. Uncertainty associated with data available to EPA for stream characteristics such as storm flows, water quality, or the physical attributes of the stream can change how the CAAP effluent will affect receiving water quality. These changes in stream characteristics should result in a systematic error (either up or down in terms of changes to water quality) that should not impact the relative differences associated with the regulated CAAP effluent. However, if the stream characteristics are too different than actual conditions (e.g., flows are much greater than modeled or background water quality masks any changes or influence by the facility) then the differences between baseline and regulated conditions may be masked.

  The results of running the QUAL2E simulations of the HHH extrapolation category with Case Studies # 2 and 3 show how changes in facility flow or effluent concentration will alter the results of the changes in water quality. For example, in the original Case Study

#2 model, BOD first decreases and remains relatively constant 30 km downstream as it begin to slightly recover. The results of the adjusted Case Study #3 show BOD increasing first and then decreasing going downstream. This example show how differences in facility and environmental conditions used for flow and effluent concentrations may result in differences in absolute values of baseline water quality for different QUAL2E simulations. Again, it is important to note that the simulated *change* in water quality is the variable of most interest in the benefits analysis.

■   Uncertainty in model specification. QUAL2E, like other water quality models, simulates certain physical, chemical, and biological processes. However, in many cases specific parameters must be estimated when actual values are not available. For example, mean solar radiation values for a facility were based on an average of mean daily solar radiation values for different cities in the state where the facility is located. Rate coefficients for nitrogen transformations, such as nitrogen hydrolysis and nitrite oxidation, are based on a set of typical ranges. Modelers often choose the average value of a range when no other data is available. When no specific SOD monitoring data was available for the modeled streams, the average SOD rate for a sandy bottom river was used to represent the real stream being modeled.

■   Uncertainty in survey weights. As described in section 8.2.2, EPA established survey weights based on facility type, predominant species, and predominant production system type. In applying these survey weights to develop a national benefit estimate from the detailed questionnaire facilities in a particular extrapolation category, EPA assumed that the facilities represented by the survey weights would experience similar benefits as the facilities in each extrapolation category. However, the facilities that are represented by the survey weights are not necessarily similar to the facilities in each extrapolation category with respect to important factors that drive water quality responses (e.g., percent load reduction resulting from the regulation; baseline pollutant concentration in effluent, and dilution ratio) and important factors that drive benefits (e.g., state populations). A better method of extrapolation would involve estimating the number of facilities in the total in-scope population that are similar to the detailed questionnaire facilities on the basis of key factors that drive water quality benefits. The use of the survey weights for extrapolating leads to additional uncertainty in EPA's national benefits estimate.

8.3   REFERENCES

Brown, L.C., and T.O. Barnwell, Jr. 1987. *The Enhanced Stream Water Quality Models QUAL2E and QUAL2E-UNCAS: Documentation and User Manual*. EPA 600-3-87-007. U.S. Environmental Protection Agency, Office of Research and Development, Environmental Research Laboratory, Athens, GA.

Carson, R.T., and R.C. Mitchell. 1993. The value of clean water: the public's willingness to pay for boatable, fishable, swimmable quality water. *Water Resources Research* 29: 2445-2454.

Hochheimer, J. 2004. *Site Selection for Water Quality Modeling*. Tetra Tech, Inc., Fairfax, VA.

Hochheimer, J., D. Mosso, and J. Harcum. 2004a. Final: Case Study 1 QUAL2E Model. Tetra Tech, Inc., Fairfax, VA.

Hochheimer, J., D. Mosso, and J. Harcum. 2004b. *Final: Case Study 2 QUAL2E Model.* Tetra Tech, Inc., Fairfax, VA.

Hochheimer, J., D. Mosso, and J. Harcum. 2004c. *Final: Case Study 3 QUAL2E Model.* Tetra Tech, Inc., Fairfax, VA.

Hochheimer, J., D. Mosso, and J. Harcum. 2004d. *Final: Case Study 4 QUAL2E Model.* Tetra Tech, Inc., Fairfax, VA.

McClelland, N.I. 1974. *Water Quality Index Application in the Kansas River Basin.* EPA 907-9-74-001. U.S. Environmental Protection Agency, Kansas City, MO.

McGuire, L. 2004a. *Benefits Estimate, Final Rule.* U.S. Environmental Protection Agency, Washington, DC.

McGuire, L. 2004b. *Flow Data for Water Quality Modeling.* U.S. Environmental Protection Agency, Washington, DC.

McGuire, L. 2004c. *Regression Model.* U.S. Environmental Protection Agency, Washington, DC.

Miller, C. 2004. *Valuation Spreadsheets.* U.S. Environmental Protection Agency, Washington, DC.

Tetra Tech, Inc. 2004. *Supporting Documentation for Extrapolation.* Tetra Tech, Inc., Fairfax, VA.

USEPA (U.S. Environmental Protection Agency). 2000. *Guidelines for Preparing Economic Analyses.* EPA 240-R-00-003. U.S. Environmental Protection Agency, Washington, DC.

USEPA (U.S. Environmental Protection Agency). 2002. *Detailed Questionnaire for the Aquatic Animal Production Industry.* OMB Control No. 2040-0240. U.S. Environmental Protection Agency, Washington, DC.

USEPA (U.S. Environmental Protection Agency). 2003. *Estimation of National Economic Benefits Using the National Water Pollution Control Assessment Model to Evaluate Regulatory Options for Concentrated Animal Feeding Operations (CAFOs).* EPA-821-R-03-009. U.S. Environmental Protection Agency, Washington, DC.

USEPA (U.S. Environmental Protection Agency). 2004. *Development Document for the Final Effluent Limitations Guidelines and Standards for the Concentrated Aquatic Animal Production Point Source Category.* EPA 821-R-04-012. U.S. Environmental Protection Agency, Washington, DC.

CHAPTER 9

OTHER REGULATORY ANALYSIS REQUIREMENTS

This section addresses the requirements to comply with Executive Order (EO) 12866 and the Unfunded Mandates Reform Act (UMRA), both which require Federal agencies to assess the costs and benefits of each significant rule they propose or promulgate.

Section 9.1 describes the administrative requirements of both EO 12866 and UMRA. Section 9.2 identifies the need for and objective of the rule. Section 9.3 provides a summary of the total social costs of the final regulations. Section 9.4 presents the estimated impacts of the final rule on noncommercial facilities. Section 9.5 summarizes the estimated monetized benefits under the final regulations and provides a comparison of the estimated total social costs and benefits under alternative regulatory options considered by EPA during the development of this rulemaking. A summary is presented in Section 9.6. Much of the information provided in this section is summarized from other sections of this report.

9.1   ADDITIONAL ADMINISTRATIVE AND REGULATORY REQUIREMENTS

9.1.1   Requirements of Executive Order 12866

Under Executive Order 12866 (58 FR 51735, October 4, 1993), the Agency must determine whether a regulatory action is "significant" and therefore subject to OMB review and the requirements of the Executive Order. Executive Order 12866 defines "significant regulatory action" as one that is likely to result in a rule that may:

- have an annual effect on the economy of $100 million or more or adversely affect in a material way the economy, a sector of the economy, productivity, competition, jobs, the environment, public health or safety, or state, local, or tribal governments or communities;

- create a serious inconsistency or otherwise interfere with an action taken or planned by another agency;

- materially alter the budgetary impact of entitlements, grants, user fees, or loan programs or the rights and obligations of recipients thereof; or

- raise novel legal or policy issues arising out of legal mandates, the President's priorities, or the principles set forth in the Executive Order."

This final regulation does not meet the criterion of $100 million in annual costs for a "significant regulatory action" because the total costs of the rule are estimated to be $1.4 million (2003 pre-tax dollars). EPA, however, submitted the action to the Office of Management and Budget (OMB) for review.

9.1.2 Requirements of the Unfunded Mandates Reform Act (UMRA)

Title II of the Unfunded Mandates Reform Act of 1995 (Public Law 104-4; UMRA) establishes requirements for Federal agencies to assess the effects of their regulatory actions on State, local, and tribal governments as well as on the private sector. Under Section 202(a)(1) of UMRA, EPA must generally prepare a written statement, including a cost-benefit analysis, for proposed and final regulations that "includes any Federal mandate that may result in the expenditure by State, local, and tribal governments, in the aggregate or by the private sector" in excess of $100 million per year.[26] As a general matter, a federal mandate includes Federal Regulations that impose enforceable duties on State, local, and tribal governments, or on the private sector (Katzen, 1995). Significant regulatory actions require OMB review and the preparation of a Regulatory Impact Assessment that compares the costs and benefits of the action.

State government facilities are within the scope of the regulated community for this final regulation. EPA has determined that this rule would not contain a Federal mandate that may result in expenditures of $100 million or more for State, local, and tribal governments, in the aggregate, or the private sector in any one year. The total annual cost of this rule is estimated to be $1.4 million (2003 pre-tax dollars). Thus, the final rule is not subject to the requirements of Sections 202 and 205 of the UMRA. The facilities which are affected by the final rule are (1) direct dischargers, (2) with flow-through, recirculating, or net pen systems, (3) engaged in concentrated aquatic animal production, and (4) with annual production of more than 100,000 lbs/yr. These facilities would be subject to the requirements through the issuance or renewal of an NPDES permit either from the Federal EPA or authorized State governments. These facilities should already have NPDES permits as the Clean Water Act requires a permit be held by any point source discharger before that facility may discharge wastewater pollutants into surface waters. Therefore, the final rule could require these permits to be revised to comply with revised Federal standards, but should not require a new permit program be implemented.

EPA has determined that this rule contains no regulatory requirements that might significantly or uniquely affect small governments. EPA is not proposing to establish pretreatment standards for this point source category which are applied to indirect dischargers and overseen by Control Authorities. Local governments are frequently the pretreatment Control Authority but since this regulation proposes no pretreatment standards, there would be no impact imposed on local governments. The requirements of the final rule are not expected to impact any tribal governments, either as producers or because facilities are located on tribal lands. Thus, this final regulation is not subject to the requirements of section 203 of UMRA.

EPA, however, is responsive to all required provisions of UMRA, including:

- ▪ Section 202(a)(1)—authorizing legislation (see Section 1.1 of this report and the final rule preamble);

- ▪ Section 202(a)(2)—a qualitative and quantitative assessment of the anticipated costs and benefits of the regulation, including administration costs to state and local governments (see Sections 4 and 7 of this report, and a summary provided in this section);

---

[26] The $100 million in annual costs is the same threshold that identifies a "significant regulatory action" in Executive Order 12866.

- ▪️ Section 202(a)(3)(A)—accurate estimates of future compliance costs (as reasonably feasible; see Section 4.3);

- ▪️ Section 202(a)(3)(B)—disproportionate effects on particular regions, local communities, or segments of the private sector (as discussed in Section 5.1 of this report, EPA identified no disproportionate impacts as a result of this final regulation);

- ▪️ Section 202(a)(4)—effects on the national economy (as discussed in Section 5.2 of this report, because of the small cost associated with the rule, EPA anticipates no discernable effects on the national economy);

- ▪️ Section 205(a)—least burdensome option or explanation required (discussed in this section).

- ▪️ Section 202(a)(5) and 204—consultation with stakeholders (described in EPA's Notice of Data Availability on the proposed rule (USEPA, 2003) and the preamble to the final rulemaking, which summarize EPA's consultation with stakeholders including industry, environmental groups, states, and local governments.

9.2   NEED FOR THE REGULATION

Section 6.3 presents EPA's discussion of the need for and the objectives of this final regulation. The concerns include water quality impairment and the introduction of non-native species.

9.3   TOTAL SOCIAL COSTS

9.3.1   Costs to In-Scope Commercial and Noncommercial Facilities

In 2003 pre-tax dollars, annualized costs for all commercial and noncommercial facilities within the scope of the rule are $1.4 million, see (Table 4-3).

9.3.2   Costs to the Permitting Authority (States and Federal Governments)

NPDES permitting authorities incur administrative costs related to the development, issuance, and tracking of general or individual permits. State and Federal administrative costs to issue a general permit include costs for permit development, public notice and response to comments, and public hearings. States and EPA might also incur costs each time a facility operator applies for coverage under a general permit due to the expenses associated with a notice of intent (NOI), which include costs for initial facility inspections and annual record-keeping expenses associated with tracking NOIs. Administrative costs for an individual permit include application review by a permit writer, public notice, and response to comments. An initial facility inspection might also be necessary.

All of the aquaculture facilities in the scope of this final regulation are currently permitted, so incremental administrative costs of the regulation to the permitting authority are expected to be negligible. However, Federal and State permitting authorities will incur a burden for tasks such as reviewing and certifying the BMP plan and reports on the use of drugs and chemicals. EPA estimates these costs at approximately $13,176 for the three-year period covered by EPA's information collection request, or roughly $4,392 per year. These results show that the recordkeeping and reporting burden to the permitting authorities is less than two-tenths of one percent of the pre-tax compliance cost for the final rule.

9.3.3   Other Social Costs

An estimate of total social costs of the proposed regulations comprises costs that go beyond the compliance costs of constructing and implementing pollution control procedures. Additional monetary costs include the cost of Federal and State subsidies in the form of a tax shield (or lost tax revenue) and costs of administering a regulation (permitting costs). The first type of cost is captured through the use of the pre-tax annualized costs for the industry. For this rule, the difference between estimated pre- and post-tax costs is $79,000 per year (see table 4-3). Section 9.3.2 described EPA's estimates the second type of cost.

Other types of social costs include possible social costs of worker dislocations, if regulated facilities are projected to close as a result of this rule. These costs comprise the value to workers of avoiding unemployment and the costs of administering unemployment, including the costs of relocating workers, and the inconvenience, discomfort, and time loss associated with unemployment. (The unemployment benefits themselves are, generally, considered transfer payments, not costs).

Another potential social costs include the cost associated with a slowdown in the rate of innovation. In theory, there might be some impact on the rate of innovation to the extent that regulated aquaculture facilities might invest in newer technologies if they did not have to allocate resources to meeting the requirements of the regulations. Generally, however, unless an industry is highly technical, with major investments in research and development, impacts on the rate of innovation are likely to be minimal.

For this rule, EPA did not evaluate these other potential social costs but expects that these costs will be modest. Among commercial facilities, EPA estimates no facility closures as a result of this final regulation. Therefore, in the commercial sector, EPA expects no job losses among commercial facilities because of this rule. Among noncommercial facilities, however, EPA's analysis indicates that 4 nonncommercial facilities may be adversely affected and possibly close as a result of this rule. This could result in job losses and worker dislocation at these facilities. Because these are noncommercial entities, it is impossible for EPA to predict what type of changes will actually occur at these facilities. EPA expects no change or slowdown in the rate of innovation in this industry as a rule of this final rule, based on EPA's analysis showing no industry changes in the commercial sector.

9.4   POTENTIAL IMPACTS ON NONCOMMERCIAL FACILITIES

EPA identified 141 Federal, State, Tribal and Alaskan non-profit hatcheries within the scope of the rule. Four of these facilities incur pre-tax annualized costs of compliance that exceed 10 percent of operating budget. Although all states report having fishing license and other user fees, not all state facilities report user fees as contributing to their operating budget. None of four facilities report user fees as a source of funding for the operating budgets, hence, none of them would be able to recoup the increased costs through increased user fees.

9.5   COMPARISON OF COST AND BENEFITS ESTIMATES

Table 9-1 compares the cost of the final rule to the economic value of the environmental benefits EPA is able to monetize (i.e., evaluate in dollar terms). EPA estimates the monetized benefits of the final rule to range from $66,214 to $98,616 per year. These benefit estimates are expressed as pre-tax, 2003 dollars and have been calculated assuming a 7 percent discount rate. Monetized benefit categories are primarily in the areas of improved surface water quality (measured in terms of enhanced recreational value). EPA also identified a number of benefits categories that could not be monetized, including reductions in feed contaminants and spilled drugs and chemicals released to the environment, as well as better reporting of drug usage to permitting authorities. These benefits are described in more detail in Sections 7 and 8 of this report and other supporting documentation provided in the record.

Table 9-1
Estimated Pre-Tax Annualized Compliance Costs and Monetized Benefits

Production System	Pre-tax Annualized Cost (Thousands, 2003 dollars)
Social Cost	
Flow-through	$1,385
Recirculating	$21
Net Pen	$36
Subtotal (Industry Costs)	**$1,442**
State and Federal Permitting Authorities	$3
Estimated Total Costs	**$1,445**
Monetized Benefits	
	$66 to $99
Estimated Total Benefits	**$66 to $99**

Note: Totals may not sum due to rounding.
*Monetized benefits are not scaled to the national level.

These estimated benefits compare to EPA's estimate of the total social costs of the final regulations of $1.4 million per year. These costs include compliance costs to all regulated facilities, and administrative costs to Federal and State governments. EPA estimates the administrative cost to Federal and State governments to implement this rule is about $3 thousand per year. There may be additional social costs that have not been monetized. These benefit estimates are also expressed as pre-tax, 2001 dollars and have been calculated assuming a 7 percent discount rate. See Section 4.3 of this report for more information.

9.6   SUMMARY

Pursuant to section 205(a)(1)-(2), EPA has selected the "least costly, most cost-effective or least burdensome alternative" consistent with the requirements of the Clean Water Act (CWA) for the reasons discussed in the preamble to the rule. EPA is required under the CWA (Section 304, Best Available Technology Economically Achievable (BAT)) to set effluent limitations guidelines and standards based on BAT considering factors listed in the CWA such as age of equipment and facilities involved, and processes employed. EPA is also required under the CWA (Section 306, New Source Performance Standards (NSPS)) to set effluent limitations guidelines and standards based on Best Available Demonstrated Technology. The preamble to the final rulemaking and Section 6.3 review EPA's steps to mitigate any adverse impacts of the rule. EPA determined that the rule constitutes the least burdensome alternative consistent with the CWA.

9.7   REFERENCES

Katzen. 1995. Guidance for implementing Title II of S.I., Memorandum for the Heads of Executive Departments and Agencies from Sally Katzen, OIRA. March 31, 1995.

USEPA (U.S. Environmental Protection Agency). 2003. Effluent Limitations Guidelines and New Source Performance Standards for the Concentrated Aquatic Animal Production Point Source Category; Notice of Data Availability; Proposed Rule. 40 CFR Part 451. *Federal Register* 68:75068-75105. December 29.

APPENDIX A

CLOSURE ANALYSIS FINANCIAL TOPICS

When Congress passed the Clean Water Act [CWA, 33 U.S.C. §1251 et seq.]), it directed EPA to require industrial dischargers to meet discharge limits based on "Best Available Technology *Economically Achievable*" (BAT, emphasis added). As EPA designs a set of economic and financial analyses tailored to the industry in a given rulemaking, the Agency incorporates methods and decisions that:

- ■ reflect the current published literature and thinking on finance and economics as tailored to that industry and appropriate for the rulemaking process,

- ■ are consistent with other EPA economic and financial analyses for effluent limitations guidelines (or document recent developments in finance and economics that lead to a change in methodology),

- ■ use multiple approaches to examine different facets of the industry, requiring the design of different tests for different sectors of the regulated community.[1]

- ■ examine an industry in the same light in which it presents itself in an EPA questionnaire, industry comments, or as presented in public data.

Chapter 3 of this report describes EPA's methodology to evaluate economic impacts to regulated facilities. For commercial facilities, the primary method is a closure analysis. This analysis requires the use of a method for calculating earnings. This Appendix discusses how EPA calculates earnings, along with a detailed discussion of other interrelated topics, including unpaid labor and management, sunk costs, capital replacement, depreciation, cash flow, and net income.

A.1   UNPAID LABOR AND MANAGEMENT

EPA received comments regarding the desirability to include proxy costs for unpaid labor and management in the economic analysis. Section A.1.1 begins by reviewing the number of facilities within the scope of the regulation that report unpaid labor and management. Section A.1.2 examines an operation's financial status prior to and as a result of the rulemaking. Section A.1.3 examines data sources for a set of estimated wages should EPA decide whether to impute this cost to a facility. Section A.1.4 reports the results of the sensitivity analysis, having addressed the question of how the economic impacts change with the imputation of costs for unpaid labor and management.

---

[1]   For example, among commercial and non-commercial operations, or multiple tests within a sector, such as financial health and borrowing/credit capacity tests as well as closure analyses for commercial facilities.

A.1.1 How Many Facilities Within The Scope of the Regulation Report Unpaid Labor and/or Management?[2]

The population within the scope of the regulation are net pen, flow-through, and recirculating systems that produce at least 100,000 lbs/year. Within the scope of the regulation, EPA's survey reports 2 unweighted facilities, representing 3 weighted facilities nationwide[3] that report unpaid labor and/or management. One facility reports only unpaid management while the other reports both unpaid labor and unpaid management. In terms of financial organization, these facilities are a S Corporation/Limited Liability Corporation, and a Sole Proprietorship. Both report annual sales less than $750,000, i.e., they are small businesses as defined by the Small Business Administration.

A.1.2 Baseline Status of These Facilities

Both facilities pass the baseline discounted cash flow analysis. Neither incur impacts under any of the five options examined for the Notice of Data Availability (USEPA, 2003) or final rule.

A.1.3 Estimated Costs for Sensitivity Analysis

EPA examined several sources for possible wage estimates to use in the sensitivity analysis. These include the Federal Minimum Wage ($5.15/hr), wage estimates by the Bureau of Labor Statistics (BLS) "Current Population Survey", and the USDA's Economic Research Service (ERS).

BLS' Current Population Survey lists median weekly earnings of full-time wage and salary workers by detailed occupation (BLS, 2004a and 2001, Table 39). For farm workers, median weekly earnings range from $309 in 2000 to $318 in 2002 or, roughly, $16,000 to $16,500 per year. For farm managers, median weekly earnings range from $547 in 2000 to $488 in 2002 or, roughly, $28,450 to $25,376 per year.

BLS' Occupational Employment and Wages for category 11-9011 Farm, Ranch, and Other Agricultural Managers in animal production reports an average (not median) annual wage of $51,370 per year in 2002 (BLS, 2004b).

As part of its Agricultural Resource Management Survey (ARMS), USDA's ERS reports an average farm household income of $65,757 per year in 2002 for all farms. For commercial farms with more than $250,000 per year in sales, farming contributes the major part of total estimated farm income at about $75,000 per year (USDA, 2003).

---

[2] No Survey IDentification Numbers (SIDs) or other identifying information are included in order to keep the report non-confidential.

[3] If the scope of the final regulation where to include all operations with more than 20,000 lbs/yr, the number of facilities reporting unpaid labor and management increases to 12 unweighted facilities and 44 weighted.

For the purposes of this analysis, EPA examined the effect of including the following labor and management costs:

■ lowest of the USDA ARMS estimate for commercial farms, USDA ARMS estimate for all farms, and Bureau of Labor Statistics, Occupational Employment and Wages ($51,370 per year).

■ 2002 Bureau of Labor Statistics, Current Population Statistics, farm manager ($25,376 per year).

■ minimum wage ($10,712 per year).

These costs were prorated according to the number of hours worked, if the respondent reported less than 40 hours/week.

A.1.4 Results of the Sensitivity Analysis

The closure analysis is based on the discounted cash flow estimate for earnings. The results of this sensitivity analysis indicate the following:

■ Under the first assumption ($51,370 per year), all facilities are baseline closures.

■ Under the second assumption ($25,376 per year), all facilities remain open in the baseline and under all of the options.

■ Under the minimum wage assumption ($10,712 per year), all facilities remain open in the baseline and under all of the options

By setting the scope of the rule to a threshold production of 100,000 lbs/yr, nearly all facilities reporting unpaid labor and management were removed from the scope. Of the three facilities that remain, none show a change in the impacts of the rule when a wage is imputed for unpaid labor and management (i.e., they are open in the baseline and remain open under all options, or they close in the baseline).

There are two issues to consider when applying charges for unpaid labor. First, the Farm Financial Standards Council specifically recommends that a "charge for unpaid family labor and management should not be included on the income statement..." (FFSC, 1997, pp. II-3 and II-22). Second, unpaid family labor is "unpaid" only with respect to the income statement. Distributions from the business to cover family living and other personal expenses are generally referred to as "family living withdrawals" or "owner withdrawals." These withdrawals are show in the statement of owner equity in the balance sheet and not the income statement.

EPA therefore does not impute a charge for unpaid labor and management when calculating farm income as cash flow or net income for the closure analysis. For the farm financial health analysis, withdrawals for family labor and management are reflected in the balance sheet information incorporated in the calculations.

A.2 SUNK COSTS

EPA received comments stating that the analysis should consider sunk costs. Comments characterized cash flow analysis as being inappropriate because it does not account for sunk costs, particularly in older facilities. Sunk costs paid out of capital (i.e., not financed) have already occurred and, as a consequence, are not incremental cash flows and should not affect future investment or the economic viability of the firm. EPA thus excludes this category of sunk costs from the closure analysis. In doing so, EPA follows standard financial textbook methodology (e.g., Brigham and Gapenski, 1997, p. 431). The Farm Financial Standards Council (FFSC, 1997) makes no mention of sunk costs.

If not expensed and financed by debt, sunk costs appear as interest and principle payments in the income statement and balance sheet. The current portion of financed sunk costs is reflected in the income statement and, thus, is included in the estimate of cash flow. The principle payment is a shift from the liabilities side to the asset side of the balance sheet. EPA considers sunk costs as reflected in a farm's debt/asset ratio and, as such, will be considered in EPA's evaluation of farm financial health and the ability of facilities (or companies) to carry additional debt (Section 3.2.4). In other words, EPA considers sunk costs as part of its economic and financial analysis.

For comparison, Engle et al. (2004) examines the potential impact of added costs on flow-through trout systems. Presumably the authors include sunk costs in their enterprise budget analysis by the inclusion of depreciation as a cost. Depreciation (as calculated for tax purposes), however, can overstate the replacement cost particularly in the initial years of an accelerated cost recovery schedule (see Section A.4). Another facet of their analysis—the mixed integer programming analysis—excludes fixed costs as well as sunk costs.

A.3 CAPITAL REPLACEMENT

EPA received comments that the facility financial analysis should include an allowance for capital replacement. EPA considered the need to include capital replacement costs in its analysis. Under the "no growth" assumption for the economic analysis, capital expenditures for growth are excluded. That is, if EPA were to include consider capital expenditures, it would be for existing assets. These expenditures fall into two categories:

- costs incurred within the useful life of the asset to keep it operating efficiently.
- costs to replace the asset when it has reached the end of its useful life.

These costs are examined in Sections A.3.1 and A.3.2, respectively.

A.3.1 Expenditures During the Useful Life of the Asset

IRS considers expenses that keep property in efficient operating condition and do not prolong the useful life or increase the capacity (i.e., add to its value as an asset) are generally deductible as repairs (CCH, 1999, p. 262, Section 903), i.e., the maintenance part of operating and maintenance costs.

This interpretation is consistent with IRS guidance to farmers and sole proprietors. For example, regarding "Instructions for Schedule F, Profit or Loss from Farming" on "Repairs and Maintenance":

You can deduct most expenses for the repair and maintenance of your farm property. Common items of repair and maintenance are repainting, replacing shingles and supports on farm buildings, and minor overhauls of trucks, tractors and other farm machinery. *However, repairs to, or overhauls of, depreciable property that substantially prolong the life of the property, increase its value, or adapt it to a different use are capital expenditures.* (emphasis added) For example, if you repair the barn roof, the cost is deductible. But if you replace the roof, it is a capital expense." (IRS, 2000, p. 25)

Regarding "Instructions for Schedule C, Profit or Loss from Business" on Line 21:

Deduct the cost of repairs and maintenance. Include labor, supplies, and other items that do not add to the value or increase the life of the property . . . Do not deduct amounts spent to restore or replace property; they must be capitalized." (IRS, 2001, p. C-4)

Capital replenishment costs within the useful life of the equipment are part of the O&M costs to keep the equipment running efficiently throughout its useful life. These costs are included in EPA's estimated compliance costs for the 10-year period. These expenses would be reported as part of Question C6 in the detailed questionnaire, as part of total expenses and, if reported as a separate cost element, as item C6.l (repairs and maintenance). Hence, EPA believes that no adjustment is needed for this component

A.3.2   Expenditures at the End of an Asset's Useful Life

The remaining scenario to examine is what happens when a major asset[4] has reached the end of its useful life. IRS states that an expense that adds to the value or useful life of property is a capital expense (RIA, 1999, §1.263(a)-1). If a major piece of equipment becomes worn down, the company would perform a discounted cash flow or other analysis to evaluate whether it makes sense to make the new investment. In that case, it is likely that a company would take the opportunity to invest in a more efficient or larger capacity item. An argument, however, could still be made that some portion of the new asset is for replacement while the remainder is for growth. The asset, however, cannot physically be apportioned and a company either installs it or not. If the asset has reached the end of its useful life, that asset plays no role in the analyses to evaluate the investment in a new asset. If the new asset is not purchased, production and revenues are zero because no production can occur without the purchase. Thus, the incremental basis for evaluating the investment is all production and all revenues, even though part of the new investment is to replace exhausted existing capacity.

Assuming the investment is made, the new costs could be financed from working capital or through debt. Each method would appear on the financial statement in a different place. If the investment is made from working capital, the asset represents a shift from current assets (cash) to fixed

---

[4] If the asset isn't major, its purchase would have no material impact in the income statement and therefore need not be considered in this discussion.

assets, i.e., no change to total assets. No adjustment is necessary to EPA's methodology if the investment is made through working capital.[5]

If the investment is financed through debt, the cost includes interest and principal. EPA's economic analyses use net income plus depreciation as the basis for cash flow with interest payments included as an expense (if interest is passed back to the facility). The current liabilities entry on the balance sheet contains the current portion of long-term debt, i.e., the principal payment due at that time. But the "no growth" assumption also implies no change in working capital. In effect, the company stops paying principal on the exhausted asset and begins paying principal on the new asset.[6] Hence, no adjustments is needed to EPA's methodology if the investment is funded through debt.

A.3.3   EPA's Consideration of Capital Replacement in the Financial and Economic Analysis

First, EPA evaluated data on capital expenditures and capital replacement. The Census Bureau collects data on annual capital expenditures including forestry, fishing, and agricultural services (Census, 2004). However, Census' capital expenditure data include intra-company transfers of capital equipment and ownership changes (Census, 2004, Appendix D-10, Instructions, Definitions, and Codes List). As a consequence, it is difficult to know whether capital expenditures help maintain existing production or whether they support expanded production. Capital expenditures for an industry undergoing consolidation, such as salmon, include acquisitions reflecting transfers of capital rather than purchases of new or replacement capital. Further, the Census data includes expansion in productive capacity, whether in new plants or in existing plants. Aggregate industry data on capital expenditures cannot be used to specify the level of capital expenditure that is necessary to maintain productive capacity at an individual facility.

Second, EPA evaluated whether depreciation represented an approximate proxy for capital replacement costs. This is discussed in Section A.4.

Third, EPA included costs for capital replacement as they occur within interest payments reported on income statement. Capital replacement costs that are capitalized and not expensed are reflected in the asset, debt, and equity components of the balance sheet as appropriate. Past capital replacement costs are represented in EPA's analysis in its consideration of farm financial health measures and credit tests that are based on balance sheet data.

Finally, when estimating compliance costs, EPA includes replacement costs for pollution control capital. EPA's cost estimates include all capital expenditures (whether initial or replacement) and O&M costs that are projected to occur within the 10-year analytical time frame.

---

[5] EPA presumes that the company included its opportunity cost of capital in the analysis to determine whether and, if so, how to fund the investment. EPA's cost annualization model includes cost of capital as an input regardless of the financial source (e.g., opportunity, debt, equity, or a mix). See Section 3.1 of this report.

[6] If the argument is made that the loan period is shorter than the useful life of the asset, the company has the benefit of using the asset when it paid off. No allowance for this benefit is made in EPA methodology.

A.4 DEPRECIATION

Depreciation is an annual allowance for the exhaustion, wear, and tear of a firm's fixed assets. Depreciation reflects expenses incurred in a prior year (i.e., sunk costs) and does not absorb incoming revenues in the current period. Depreciation may be consideredas recouping part of an expense made in a previous period (i.e., looking backward to the original purchase) or as saving toward to replacement of that asset (i.e., looking forward to the replacement purchase). The second approach assumes that the annual operation and maintenance charge is not sufficient to ensure a facility's efficiency and capacity in the long run. Over the long term, ongoing reinvestment in plant and equipment is necessary.

EPA examined the relationship between depreciation as a concept and depreciation as recorded for tax purposes to evaluate whether depreciation could serve as a proxy for capital replacement. Although depreciation is supposed to reflect wear and tear over the useful life of an asset, it does not necessary do so for tax purposes. There are several reasons why depreciation for tax purposes might bear no relationship to capital replenishment costs. First, rather than depreciating an asset over its useful life, it is depreciated over the shorter class life. For example, municipal wastewater treatment plants have a class life or useful life of 20 years or more but less than 25 years. Its recovery period for depreciation is 15 years (CCH 1999, Section 1240). There is a five-year period at the end where the company has recovered the value of the asset but does not have to replace it.

Second, a company may use the Modified Accelerated Cost Recovery System (MACRS) rather than straight-line depreciation for additional tax benefit. MACRS provides substantial tax benefits by allowing larger reductions in taxable income in the years immediately following an investment when the time value of money is greater. In our example of the 20-year wastewater treatment plant, a depreciable fraction over its useful life would be 1/20 or 0.05 (full year convention). Under MACRS, however, that fraction would be 0.10 for the first year. The effect becomes more pronounced with shorter recovery periods. The effect of different depreciation methods on earnings and the overstatement of the true economic cost of depreciation (as noted in FFSC, 1997, p. II-30). Rappaport (1998, p.14) notes that the choice of an accounting method is a management choice that can materially impact earnings but does not change a company's cash flows. Damodaran (2001, Chapter 3, p. 6) a notes that many companies legally keep two sets of books, one recording straight line depreciation for financial reporting and the other recording accelerated depreciation for tax purposes.

Third, in the scenario of a new or heavily upgraded site, the depreciation is highest when there is the least need for capital replenishment. With accelerated depreciation, the write-offs are highest during the first few years of operation when there is little need to replenish equipment. Fourth, the original cost of an asset might bear little resemblance to the replacement cost for the asset.

In theory, the economic description of depreciation as a means of prorating a capital cost over all the units it produces during its useful life is a cost that is part of the "cost of production." In practice, EPA found that depreciation as recorded for tax purposes could substantially overestimate the replacement cost for capital investments and was thus not appropriate to include as a cost in the earnings estimates. The exclusion of depreciation as a cost is consistent with economic theory that a facility will continue operation as long as price exceeds its variable costs. Depreciation is a bookkeeping charge reflecting previous capital expenditures and thus is a sunk cost which should be ignored in the closure analysis.

A.5   CASH FLOW

EPA's closure analysis is a discounted cash flow analysis (DCF). DCF methods are used in valuing companies (e.g., to decide whether to invest in a business) and projects (e.g., capital budgeting). EPA uses DCF to value a facility prior the incurrence of any additional pollution control costs. EPA then values the facility after the inclusion of these costs and compares the results. EPA's focus is on the change created by the incremental pollution control costs rather than the baseline value itself.

EPA examined several textbooks and references to identify the most appropriate basis for evaluating earnings as part of its closure methodology. Table A-1 summarizes several academic opinions on cash flow versus accounting profits as a measure on which to base the evaluation of a project or firm. There is a consensus that cash flow is the appropriate measure for the analysis. EPA's methodology, then, is consistent with those in current academic literature.

Table A-1
Cash Flow Versus Accounting Income

Source	Comment
FFSC, 1997, p. II-15	"...The most common valuation methods are: ...*Discounted Cash Flow Methods*. For an asset: the present value of future cash inflows into which an asset is expected to be converted in the due course of business, less present values of cash outflows necessary to obtain those inflows."
Brealy and Myers, 1996, pp. 113-114	Chapter Heading: Making Investment Decisions with the Net Present Value Rule. "...you should always stick to three general rules: 1. Only cash flow is relevant. 2. Always estimate cash flows on an incremental basis. 3. Be consistent in your treatment of inflation. The first and most important point is that the net present value rule is stated in terms of cash flows. Cash flow is the simplest possible concept; it is just the difference between dollars received and dollars paid out. Many people nevertheless confuse cash flow with accounting profits." The authors identify depreciation as a non-cash-flow item.
p.114-116	Specifies that sunk costs should be excluded, whereas working capital requirements, opportunity costs, and incidental effects should be included in the analysis. Brealy and Myers warn analysts to be cautious when deciding whether and, if so, how to include allocated overhead costs in the analysis.
Brigham and Gapenski, 1997 p. 429-431.	"CASH FLOW VERSUS ACCOUNTING INCOME Income statements in some respect mix apples and oranges...In capital budgeting, it is critical that we base decisions strictly on cash flows, or actual dollars that flow into and out of the company..." Incremental cash flow is identified as the appropriate basis for capital budgeting purposes. Sunk costs are not included but opportunity costs, capital outlays, effects on other projects, changes in net working capital are included.

Source	Comment
Damodaran, 2001a	Chapter 9: Measuring Earnings. "...the accounting earnings for many firms bear little or no resemblance to the true earnings of the firm." Chapter 9 outlines techniques to get from accounting statements to a measure of earnings while Chapter 10 traces the path from earnings to cash flow. Cash flow is the basis on which the value of an asset/project/firm is calculated.
Damodaran, 2001b	Professor Damodaran's web site materials are succinct: returns on projects should be measured based on cash flows. To get from accounting earnings to cash flow, you add back non-cash expenses like depreciation, subtract out cash outflows which are not expensed (like capital expenditures), and incorporate changes in working capital.
Jarnagin, 1996.	Financial Accounting Standards Board, SFAS Nos. 105, 107, and 119 state that one of the preferred methods for calculating fair value is the present value of future cash flows. The present value of future net income is not listed among the preferred methods.
McKinsey & Company, Inc., et alia, 2000	"First, empirical research suggests that cash flow, not accounting earnings, is what drives share price performance." p. 55 The authors' opinion is reflected in their title for Chapter 5: Cash is King.
Rappaport, 1998 pp.13-31.	Chapter 2: Shortcomings of Accounting Numbers "Remember, cash is a fact, profit is an opinion." (p. 15) Cash flow, not earnings or net income, is appropriate basis for valuation.

Some authors use the term "free cash flow" is used to describe the after-tax cash flow that would be available to both creditors and shareholders if the company had no debt (McKinsey & Company, Inc., 2000; Brealy and Myers, 1996; Brigham and Gapenski, 1997), although there is no consensus on how to calculate it. In general, most approaches calculate free cash flow from net income with the following components:

Free cash flow =	net income	
	plus	non-cash expenses (e.g., depreciation)
	minus	cash outflows that are not expensed (e.g., capital expenditures)
	plus/minus	changes in working capital (to change accrual revenues and expenses into cash revenues and expenses)

(See Damodaran, 2001b, slides 158 and 171; Brealy and Myers, 1996, p. 121; and Rappaport, 1998, pp. 15-18). The differences among the authors lie in the level of detail pursued when making estimates of free cash flow. Some of the factors listed in the literature include adjustments for:

- non-operating income and expenses (e.g., cash flows from discontinued operations, extraordinary gains or losses, or cash flows from investments in unrelated subsidiaries);

- lease expenses;

- R&D expenses (which Damodaran, 2001a, argues should be capitalized);

- ☐ increase in assets, net of liabilities;

- ☐ investment in goodwill (expenditures to acquire other companies in excess of book value of net assets).

"Free cash flow" might be a closer estimate to what EPA would want to examine in the closure analysis but some factors render it inappropriate for implementation. EPA takes the position that it examines a firm or facility on the basis on which it has chosen to present itself. The potentially substantial adjustments needed to estimate free cash flow from financial statements means EPA would have to justify changes in accounting practices from those used by the company. Second, in order to be consistent with its "no growth" assumption (to avoid a facility "growing" out of impacts from incremental pollution control costs), EPA would have to be able to disaggregate capital expenditures to isolate those used for existing assets from those for new capacity, mergers, or acquisitions. As mentioned in Section A.3.3, EPA's analysis addresses capital replacement considerations.

A.6   NET INCOME

EPA received comments that depreciation is a cost that should be included in the earnings estimate, that is, the basis for earnings should be net income rather than cash flow.[7] The Financial Accounting Standards Board launched a "financial performance reporting" project in 2001 (FASB, 2004). Its summary of user interviews contains two items relative to this discussion:

- ☐ "Net income is an important measure that is often used as a starting point for analysis but generally not the most important measure used in assessing the performance of an enterprise..."

- ☐ "Key financial measures include the following, which are not necessarily well-defined terms or notions: (a) 'operating' free cash flow or free cash flow, (b) return on invested capital, and (c) 'adjusted,' 'normalized,' or 'operating' earnings.

Net income, then, is not considered a key financial measure by the user community.

Part of the discussion on whether net income or cash flow should be used as earnings in the closure analysis hinges on whether the analyst evaluates returns to the firm or returns to the stockholder. As Darmodaran (2001a, Chapter 9) and Brealy and Myers (1996, pp. 766-768) note, returns to the firm begin with after-tax operating earnings while returns to stockholders begin with net income. Closure is the most serious impact that can occur at the facility level and EPA therefore considers the more conservative approach of evaluating the impacts of incremental pollution control costs as evaluating the change in returns to the firm.

Net income considers depreciation a cost but it is a non-cash cost. A company is not obliged to set aside or save the value of depreciation for capital replenishment. A company has the option of dispersing the cash represented by depreciation as it sees fit, including dispersing it as dividends to

---

[7] Section A.4 reviews the difference between the economic definition of depreciation and depreciation as calculated for tax and reporting purposes. Section A.3 reviews how EPA addresses capital replacement costs in its economic analysis.

stockholders. The ability of a company to distribute the cash represented by depreciation is what makes depreciation appropriate to consider as earnings. If a company chooses not to save towards reinvestment, that is not an impact of the rule. Site visits and survey financial data indicate that some facilities are run until worn out and the companies have not set aside the value of accumulated depreciation for replacing the equipment. Furthermore, all other things being equal, when it is time to reinvest and the firm has not set aside for that investment, the firm is in no worse shape that it was when it made the initial investment.

EPA evaluated the effects of changing from a cash flow basis to a net income basis. First, the change results in a smaller number of facilities that can be analyzed for impacts. Marginal firms who are "living off their depreciation" and remain open under the baseline cash flow analysis become baseline closures under the net income assumption. These marginal firms have the potential to be removed from the population on which EPA can evaluate impacts of incremental costs. In EPA's economic analysis for the industry, two additional facilities were projected to close as a result of the rule when net income was used as the basis for earning.

Second, neither the costs nor the removals for baseline closures are included in the analysis. This implies that the use of net income as earnings could underestimate the cost of the rule.

To avoid removing marginal facilities from the impact analysis and underestimating the cost of the rule, EPA's closure analysis uses a cash flow basis.

A.7   REFERENCES

BLS (U.S. Department of Labor. Bureau of Labor Statistics). 2004a. Current Population Survey. Table 39 Median Weekly Earnings of Full-time Wage and Salary Workers by Detailed Occupation. <http://Stats.bls.gov/cps/cpsaat39.pdf> downloaded 26 January.

BLS (U.S. Department of Labor. Bureau of Labor Statistics). 2004b. Occupational Employment and Wages. 2002 data. Category 11-9011 Farm, Ranch, and Other Agricultural Managers. http://www.bls.gov/oes/2002/oes119011.htm. downloaded 26 January.

BLS (U.S. Department of Labor. Bureau of Labor Statistics). 2001. Current Population Survey. Table 39 Median Weekly Earnings of Full-time Wage and Salary Workers by Detailed Occupation. <http://Stats.bls.gov/cps/cpsaat39.pdf> downloaded 12 April 2001.

R.A. Brealy and S.C. Myers. 1996. *Principles of Corporate Finance*. 5th edition. The McGraw-Hill Companies, Inc. New York.

Brigham, E.F., and L.C. Gapenski. 1997. *Financial Management: Theory and Practice*. 8th edition. The Dryden Press. Fort Worth, Texas.

CCH (CCH, Incorporated). 1999. *2000 U.S. Master Tax Guide*. Chicago, Illinois.

Census (U.S. Census Bureau). 2004. *Annual Capital Expenditures: 2002*. Washington, DC. ACE/02. Issued January 2004.

Damodaran, Aswath. 2001a. *Investment Valuation.* 2nd edition. John Wiley & Sons. New York. December publication date. Manuscript available at <http://www.stern.nyu.edu/~adamodar/New_Home_Page/valn2ed/book.htm> 12 December.

Damodaran, Aswath. 2001b. 2001b. *Applied Corporate Finance.* Overheads for Measuring Investment Returns and Valuation. <http://www.stern.nyu.edu/~adamodar/New_Home_Page/AppldCF> and <http://www.stern.nyu.edu/~adamodar/pdfiles/ovhds/val.pdf > downloaded 11 December 2001.

Engle, C.R., Steve Pomerleau, Fary Fornshell, Jeffery M. Hinshaw, Debra Sloan, and Skip Thompson. 2004. The Economic Impact of Proposed Effluent Treatment Options for Production of Trout *Onchorhynchus mykiss* in Flow-through Tanks. Submitted to USDA. March draft.

FASB (Financial Accounting Standards Board). 2004. Project Updates: Financial Performance Reporting by Business Enterprises. Last Updated: March 22, 2004. <www.fasb.org/project/fin_reporting.shtml> downloaded 22 March.

FFSC (Farm Financial Standards Council). 1997. *Financial Standards for Agricultural Producers.* December.

IRS (Internal Revenue Service). 2001. *2001 Instructions for Schedule C, Profit or Loss from Business.* Washington, D.C.

IRS (Internal Revenue Service). 2000. *Farmer's Tax Guide: for use in preparing 1999 returns.* Publication 225. Washington, D.C.

Jarnagin, Bill D. 1996 *Financial Accounting Standards: Explanation and Analysis.* 18th edition. CCH, Incorporated. Chicago, IL.

McKinsey & Company, Inc.. 2000. *Valuation: Measuring and Managing the Value of Companies.* , Tom Copeland, Tim Koller, and Jack Murrin. 3rd edition. John Wiley & Sons, Inc. New York.

Rappaport, Alfred. 1998. *Creating Shareholder Value: A guide for managers and investors.* The Free Press. Simon & Schuster, Inc. New York.

RIA (Research Institute of America). 1999. *The Complete Internal Revenue Code.* New York.

USDA (U.S. Department of Agriculture). 2003. *Agricultural Income and Finance Outlook.* AIS-81. Economic Research Service. November 5.

USEPA (U.S. Environmental Protection Agency). 2003. Effluent Limitations Guidelines and New Source Performance Standards for the Concentrated Aquatic Animal Production Point Source Category; Notice of Data Availability; Proposed Rule. 40 CFR Part 451. *Federal Register* 68:75068-75105. December 29.

APPENDIX B

GIS TABLES FOR CHAPTER 7

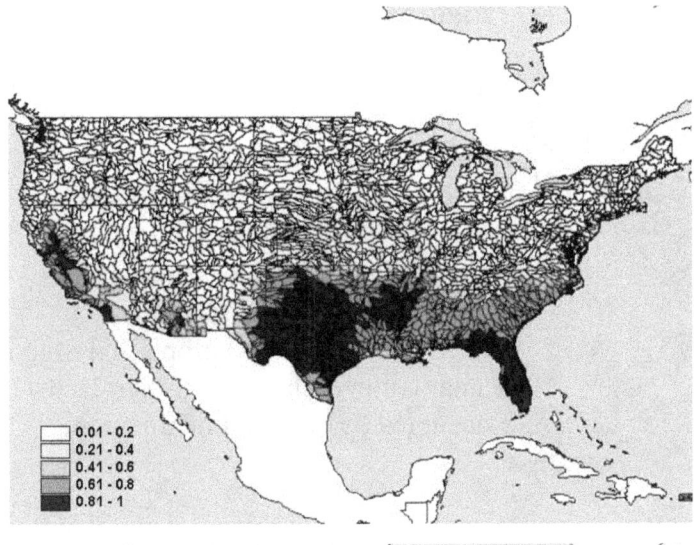

Figure 1. Potential distribution of Mozambique (A) (*Oreochromis mossambicus*), blue x Mozambique (B) (*O. aureus x mossambicus*), and Wami River x Mozambique tilapia (C) (*O. urolepis hornorum x mossambicus*) in the United States, based on known native range occurrences and 14 environmental variables, and generated using the Genetic Algorithm for Rule-set Prediction (GARP). Color scale denotes the weighted proportion of the area within each USGS 8-digit HUC occupied by potential distribution.

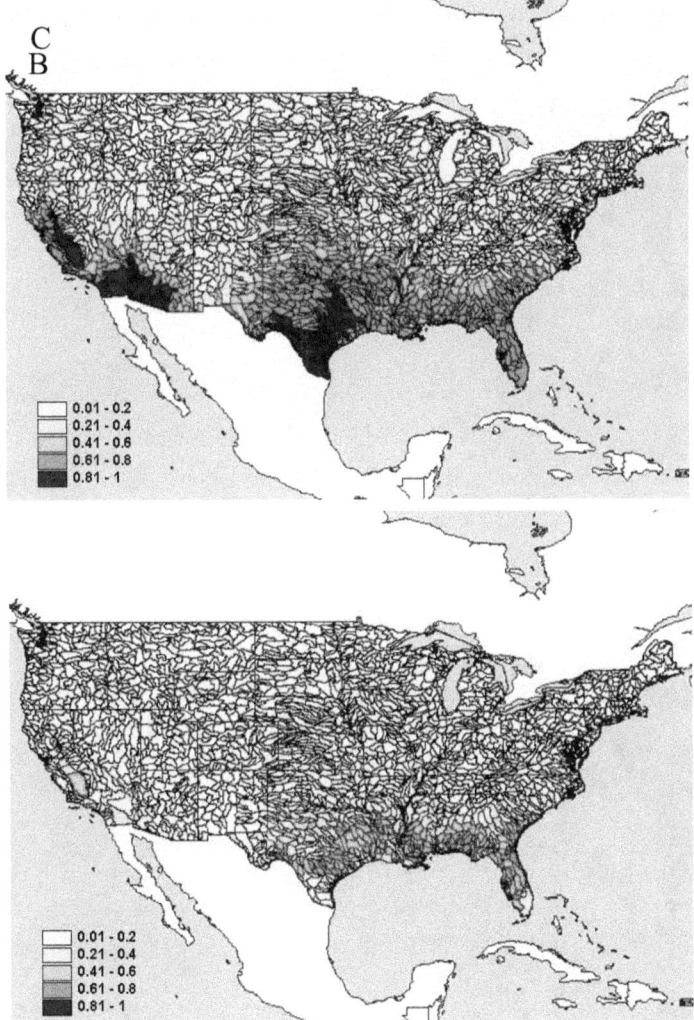

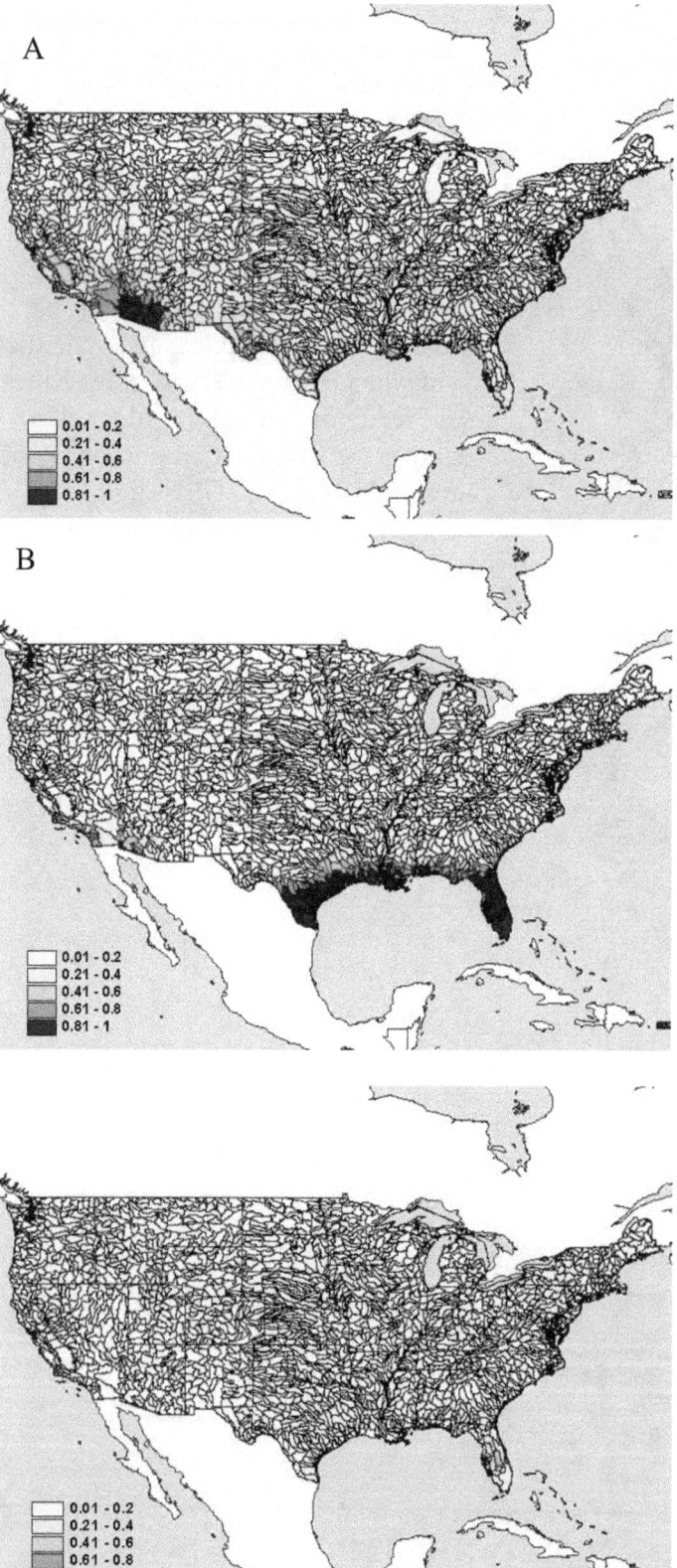

Figure 2. Potential distribution of blue (A) (*O. aureus*), Nile (B) (*O. niloticus*), and Wami River tilapia (C) (*O. urolepis hornorum*) in the United States, based on known native range occurrences and 14 environmental variables, and generated using the Genetic Algorithm for Rule-set Prediction (GARP). Color scale denotes the weighted proportion of the area within each USGS 8-digit HUC occupied by potential distribution.

www.ingramcontent.com/pod-product-compliance
Lightning Source LLC
Chambersburg PA
CBHW080636180526
45168CB00008B/3195